PORPHYRIN PHOTOSENSITIZATION

ADVANCES IN EXPERIMENTAL MEDICINE AND BIOLOGY

Recent Volumes in this Series

PORPHYRIN PHOTOSENSITIZATION

Edited by
David Kessel
Wayne State University
Detroit, Michigan

and

Thomas J. Dougherty
Roswell Park Memorial Institute
Buffalo, New York

PLENUM PRESS • NEW YORK AND LONDON

Library of Congress Cataloging in Publication Data

Porphyrin Photosensitization Workshop (1981: Washington, D.C.)
Porphyrin photosensitization.

(Advances in experimental medicine and biology; v. 160)
"Proceedings of a Porphyrin Photosensitization Workshop, held September 28-29, 1981, in Washington, D.C."—Verso t.p.
Includes bibliographical references and index.
1. Porphyrin and porphyrin compounds—Physiological effect—Congresses. 2. Photosensitization, Biological—Congresses. 3. Photochemotherapy—Congresses. I. Kessel, David. II. Dougherty, Thomas J. (Thomas John), 1933- . III. Title. IV. Series. [DNLM: 1. Prophyrins—Therapeutic use—Congresses. 2. Porphyrins—Pharmacodynamics—Congresses. 3. Photochemotherapy—Methods—Congresses. W1 AD559 v. 160/QU 110 P836p 1981]
QP671.P6P67 1981 616.99′406 82-19001
ISBN 0-306-41193-8

Proceedings of a Porphyrin Photosensitization Workshop, held September 28-29, 1981, in Washington, D.C.

A Division of Plenum Publishing Corporation
233 Spring Street, New York, N.Y. 10013

Printed in the United States of America

PREFACE

This book contains contributions from participants at a Porphyrin Photosensitization Workshop held in Washington DC on 28-29 September, 1981. There had been several prior such gatherings during 1979 and 1980, but the 1981 meeting was the most extensive to date with almost 200 attending. These periodic meetings are designed to enable workers in the field to exchange information and learn of recent developments. The Workshop was supported by PHS grant CA 31188 awarded by the National Cancer Institute, DHHS. Additional funds from Coherent Inc. [Laser Division], Palo Alto, contributed to the success of this meeting as did the many excellent presentations and discussions.

Preparation of camera-ready copy from submitted manuscripts was carried out via the Textedit computer program by Glenn Cummings, Amanda Cummings and David Kessel. Responsibility for final manuscript preparation rests with the editors.

David Kessel

Thomas J. Dougherty

CONTENTS

INTRODUCTION

David Kessel and Thomas J. Dougherty

Departments of Pharmacology and Oncology
Wayne State University School of Medicine
Detroit MI, 48201

Department of Radiation Biology
Roswell Park Memorial Institute
Buffalo, NY 14263

Like many other reports on biologic discoveries, porphyrin localization by tumor tissues has been described at 10-20 year intervals with the first report appearing in 1924. During the current cycle, appropriate instrumentation was available so that the phenomenon could be translated into clinical practice. Collaboration and exchange of information between workers in many disciplines will be required for further development of what has come to be called Photoradiation Therapy (PRT). Since clinical application of tumor localization is so recent, opportunities for such collaboration has been limited.

This Workshop was designed to bring together workers involved in fundamental studies on porphyrin chemistry, physics and photobiology, along with those using porphyrins for tumor localization and tumor eradication in animals and in man. Data relating to the important fields of light dosimetry and instrumentation for detection of porphyrin fluorescence and for photoradiation therapy was also presented.

Results reported here, along with information published elsewhere, indicate that photoradiation therapy can be useful as a means of tumor eradication or palliation. Furthermore, tumor localization procedures involving porphyrins can detect neoplastic foci where more conventional procedures were unsuccessful. It now becomes important to define the specific indications for photoradiation therapy and for use of porphyrin localizing procedures.

With regard to the tumor localization described in these studies, it must be emphasized that the product used in the clinic, HPD (hematoporphyrin derivative), is a complex mixture of porphyrins derived from hematoporphyrin by synthetic reactions. After injection into tumor-bearing animals or cancer patients, the porphyrin components of HPD are initially widely distributed. After a 24-72 hr interval, localization of fluorescence can be observed in tumor tissues along with certain normal host organs including liver, kidney and spleen. Porphyrin accumulation in skin is also observed, accounting for transient skin photosensitization.

Unanswered questions relate to identification of the active components of HPD, mechanism(s) of tumor localization and determinants of PRT cytotoxicity. Current research efforts are also being directed towards development of improved light-delivery systems, and delineation of light and drug dosimetry parameters. This publication represents a compilation of current research results along with suggestions concerning the utilization of this new and novel form of cancer therapy.

PHOTORADIATION THERAPY - CLINICAL AND DRUG ADVANCES

T.J. Dougherty, D.G. Boyle, K.R. Weishaupt,
B.A. Henderson, W.R. Potter, D.A. Bellnier
and K.E. Wityk

Roswell Park Memorial Institute
Buffalo, New York

INTRODUCTION

Photoradiation therapy (PRT) using hematoporphyrin derivative (HPD[1]) as photosensitizer is becoming widely investigated by many groups in the U.S. and abroad for a variety of malignancies. At Roswell Park, while most of our reported data deals with treatment of cutaneous and subcutaneous lesions, largely in patients with advanced disease, we are increasingly investigating PRT for other types of lesions. Elsewhere in this volume, our studies on endoscopic PRT for lung and tracheal lesions will be described by Dr. Vincent, and studies of esophageal tumors by Dr. Douglass. In addition we are carrying out treatment of brain tumors with Dr. West and head and neck tumors with Drs. Shedd, Sako and Raczak. In addition, we will be reporting on several basic studies regarding the active component of HPD as well as uptake and retention studies of HPD in vitro (Henderson, Bellnier).

Conclusions regarding Metastatic Cutaneous and Subcutaneous Tumors

In a group of 35 patients in this category, most of whom had many individual lesions, were in advanced stage of disease and had recurrent tumor following several conventional modalities, we can make the following generaliza-

---

[1] See Addendum.

tions. Patients with small, superficial lesions even if quite extensive, e.g. over one quarter to one half of the chest wall, can benefit from PRT even if the lesions occur in a prior ionizing radiation field. Patients with large lesions, if not extensive, i.e. localized, can also benefit from PRT. Patients who are unlikely to benefit are those with extensive disease (more than one half to the chest wall) or those with numerous or contiguous large lesions.

For superficial lesions optimum conditions, in our hands, are 2.5 mg/kg HPD (Day 0), Day 3-5, 54 Joules/cm² (60 mW/cm² x 15 min) of light delivered from a 200 μM or 400 μM quartz fiber (620-640 nm from the Rhodamine dye laser). This can provide complete response of such lesions without unacceptable damage to normal skin. In general on Day 2, control areas (1.5 x 1.5 cm) of normal skin are exposed to various light doses to preclude an over-reaction to treatment since not all patients, nor all sites, respond identically. For large lesions we use interstitial PRT generally following an HPD dose of 5.0 mg/kg (Day 0). On Day 3-7, treatment is carried out by inserting single or multiple fibers directly into the tumor via 18 gauge needles. As a rule of thumb (empirical observation from primary tumors in dogs and cats) we take a single fiber delivering 300 mW for 20 min to control approximately 10 cc of tumor. Therefore for large lesions fibers are inserted 1-1.5 cm apart and advanced 1-1.5 cm through the tumor every 20 min to cover the entire volume. The power per fiber is limited to 300-400 mW to prevent thermal effects at the fiber tip. We are currently investigating use of a diffusing surface on the fiber to increase the power on a single fiber, and to treat a larger volume per fiber.

Complications which we have seen include inflammatory breast cancer where a much enhanced reaction is seen when the optimum light dose is used. Considerable skin necrosis was seen under these conditions. While healing occurred eventually and good tumor control was obtained, there was considerable pain involved; if a large area had been treated the result could have been disastrous. An enhanced reaction was also seen in patients who received adriamycin within a few months of PRT, although this reaction was limited to the area of prior radiotherapy only. Similar over-reaction was seen in inflamed areas from hyperthermia.

Patient follow-up has ranged to four years (1 patient) without recurrence. Of the 35 patients discussed above we conclude that only 5 actually benefited from PRT treatment. However only 12 patients of the 35 were in sufficiently good condition to have been expected to benefit from treatment of any kind. Patients with cutaneous (superficial) tumors clearly respond best to treatment. Patients with large, wide-spread cutaneous lesions respond to treatment but often experience too much pain due to extensive tumor necrosis to complete therapy.

Interstitial PRT

Patients with localized, large cutaneous lesions can be effectively treated using interstitial PRT. Lesions as large as 5 x 5 cm have been completely eradicated using this method. In general the patients receive 5.0 mg/kg HPD, 3 days prior to light treatment. Two or more fibers, each delivering approximately 300 mW are inserted simultaneously, directly into the lesion (generally without local anesthesia) through 18 gauge needles as described above.

These techniques were first worked out in primary tumors in pet cats and dogs in a collaborative project with Richard Thoma, D.V.M. To date, more than 45 solid tumors have been treated including 13 histological types. Complete responses currently are at 43%, with 30% controlled one year or longer (up to 4 yeas) including osteosarcoma (head and neck area), melanoma, adenocarcinoma, squamous cell carcinoma and fibrosarcoma. We have had a complete response in an adamantinoma which demonstrated a very high degree of HPD fluorescence although this was considered to be a benign tumor. Most of the tumors showing complete responses were 4.0 cm or less in size. We rarely see tumors this small in humans. Clearly, PRT can eradicate solid tumors permanently, provided an adequate light dose is delivered to the entire volume.

In order to accomplish many of these treatments, we have designed and built beam splitting devices and specialty fibers as required. Fibers which allow for a cyclindrical pattern are most useful in treating obstructing and other tumors in the bronchus, trachea and esophagus. The effectiveness of PRT for controlling early lung cancers is most notable in the work of Hayata and Kato reported at this meeting.

HPD Fractionation

HPD is a mixture of porphyrins derived from hematoporphyrin by an acetic acid - sulfuric acid treatment followed by reaction in basic medium. The identification of the particular porphyrin or porphyrins responsible for the biological responses is unknown. We report here the isolation and biological efficacy of eight components of HPD, isolated by reverse phase HPLC or by gel filtration chromatography. The biologically active fraction has been isolated and partially characterized.

MATERIALS AND METHODS

Hematoporphyrin derivative (HPD) was prepared as previously described by us (1,2) from hematoporphyrin hydrochloride (Roussel, New York NY). Protoporphyrin and hydroxyvinyldeuteroporphyrin were obtained from porphyrin Products (Logan, UT).

Chromatography

Thin layer chromatography (TLC) was carried out on silica plates (Polygram SIL-N-HR, Brinkman Instruments, Westbury, NY) with benzene, water (65:35:1.5). High performance liquid chromatography (HPLC) was performed using a µBondpak C-18 column (3.9 mm x 30 cm) (Waters Associates, Milford, MA) with variable wavelength detector (400 or 500 nm generally). Samples were eluted with methanol, water, acetic acid (80:20:4). For obtaining samples for biologic activity the eluting solvent was THF, water (80:20). Peak areas on HPLC scans were determined with a planimeter. Gel filtration was carried out with Bio-Gel (Bio-Rad, Richmond, CA) of various exclusion limits ranging from 2000 (Bio-Gel P2) to 1.5×10^6 (Bio-Gel A-1.5m). Bio-Gel P-10 (exclusion limit 20,000) was used most frequently.

Animal Tumor System

The SMT-F tumor was implanted subcutaneously by trocar into the axillary region of DBA/2 HeLa mice (4). Tumors grew to 5-7 mm in diameter in 5-7 days, at which time treatment was initiated.

Treatment

Mice with 5-7 mm diameter tumors were injected intraperitoneally with various porphyrins or porphyrin fractions in doses of 2.25 to 7.5 mg/kg body weight. Porphyrin concentrations in 0.9% saline were determined by measuring the optical density at 500 nm. This was found to be more accurate than weight for many of the smaller fractions. Comparisons are therefore on a molar rather than on a weight basis. The day following injection, the area over the tumor was shaved with a small margin, and the animals were restrained in holders without anesthesia. The holders had holes through which the tumors could be exposed to the activating light which was at 600-700 nm, and delivered at a dose rate of 160 mw/cm² for 30 min (288 Joules/cm²). The entire tumor and 2-3 mm margin were exposed. The lamp was a 1 KW xenon lamp (Oriel Optical Corp., Stamford, CN) filtered with water (3 in), an IR reflecting mirror (Baird Atomic) and a red cut-off filter (Corning C52-62). Tumor response was judged starting the day after treatments. Non-palpable tumors were considered as a complete response. The time in days for half of the entire group of animals to demonstrate tumor regrowth was taken as TC_{50}. With HPD at an injected dose of 7.5 mg/kg the TC_{50} is 7-10 days and was consistent in more than 10 separate tests of 8-10 mice per test group. Initial complete responses (i.e. Day 1) were seen in 90-100% of the animals. Ultimately 80-100% of the tumors regrew within 90 days. Using 10 mg/kg HPD, 30% of the tumors were controlled for 90 days or longer.

RESULTS

The major components of HPD, identified by HPLC, are hematoporphyrin (HP), 42% and hydroxyethylvinyldeuteroporphyrin (HVD), 34%. Identity was by comparison with retention times of authentic samples in HPLC and TLC. HP was resolved into diastereoisomers and HVD into structural isomers by HPLC. In addition to the major components, HPLC reveals six minor components comprising the remainder of the material. Each of these components was tested for biologic activity as described above following isolation from the HPLC column, after the solvent was removed by evaporation. Results are summarized in Table 1. Tumors in animals not receiving porphyrin showed no response to the light alone. The standard for comparison was HPD itself which caused 80% initial complete response of the SMT-F tumor with 50% of the tumors controlled at 7-10 days post treatment (TC_{50}). Other samples were scored according to the drug dose required to achieve a similar response, all other variables remaining constant (TC_{50} dose). It was readily apparent that the major components, HP and HVD at 7.5 mg/kg caused no tumor response under these conditions. Since our intent was to identify the component(s) of HPD more active than HPD itself, rather than to quantitate relative activities, we did not determine the drug dose necessary for inactive materials to achieve results equivalent to HPD. We also determined efficacy of protoporphyrin, since it has been identified as a component of HPD previously. In our hands, it co-migrated with component 8 by HPLC and could not be specifically identified. However, protoporphyrin was not active by this test. Protoporphyrin was also injected in a more concentrated solution, i.e. 5.0 mg/ml, to determine the effect of aggregation state. This also tested negative.

We also studied the isolation of the HPD components by gel filtration using aqueous conditions under which non-covalent polymers might be expected to be more stable. With various Bio-Rad gel columns, HPD (5 mg/ml) is readily separable into several fractions immediately distinguishable by color. Using a P-10 column with a nominal exclusion limit of 20,000 three distinct regions are observable. The most rapidly moving fraction [A], dark brown in color, comprised 40-50% of the mixture. The broad, diffuse red-brown band

Table 1. Biologic Activity of HPD Components

Component†	Identity	Isolation Method	TC_{50} Dose (mg/kg)‡
HPD (100%)	Mixture	-	7.5
1 (41.7%)	HP	HPLC*	>7.5
2 (4.6%)	-	HPLC	-
3 (34.1%)	HVD	HPLC	>7.5
4 (2.7%)	-	HPLC	3.75
5 (2.7%)	-	HPLC	3.75
6 (0.9%)	-	HPLC	3.75
7 (1.14%)	-	HPLC	3.75
8 (12.1%)	-	HPLC	3.75
4-8 (19.5%)	-	HPLC	2.5-3.75
A (40-50%)	Porphyrin aggregates (HVD+?)	GF*	3.75-4.2
B (10-25%)	HP aggregates	GF	>7.5
C (25-40%)	HP, HVD, PP	GF	>7.5
PP*	-	-	>7.5
HP*	-	-	>7.5
HVD*	-	-	>7.5

†Number in parenthesis are percentages of total mixture.

‡Drug dose required to achieve 50% complete response at 7-10 days post treatment as described in Materials and Methods; >7.5 mg/kg indicates that at this drug dose there were 0% complete responses from 0 to 5 days post treatment at which time all tumors were growing rapidly as in controls.

*HPLC=high-pressure liquid chromatography; GF=gel filtration; PP (protoporphyrin) and HP (hematoporphyrin) were commercial products, HVD was isolated via HPLC.

[**B**] eluted just after **A**, constituted 25% of the mixture. The final dark red, slowly eluted fraction [**C**] comprised 28%. Essentially quantitative recovery from the column was obtained. When a more dilute HPD solution (0.5 mg/ml) was analyzed, we observed an increase in fraction **C** and decrease in **B**. All biologic activity resided in **A** which was 1.5-2.0 more active than was HPD, i.e. an injected dose of 3.75 to 4.2 mg/kg of **A** yielded the same TC_{50} as 7.5 mg/kg HPD. Silica TLC studies of each of these fractions indicated fraction **A** mainly remained at the origin (Rf=0-0.01), but also contained small amount of a material that co-migrated with HVD (Rf=0.2) with a minor portion migrating just above the origin Rf=0.03-0.06. Fraction **C** contained only HP, HVD and protoporphyrin (Rf=0.39), with HP > HVD > protoporphyrin (qualitatively, based on relative fluorescence). When fraction **A** was again passed through the gel column, it migrated as a single, brown band even when further diluted by a factor of 10. However, fraction **B** separated into **B** and **C** depending on relative dilution. Fraction **C** was unchanged migrating as a relatively tight red band. HPLC analysis of fraction **A** revealed HP (13.2%), HVD (45%), 4 (3.8%), 5 (9.4%), 6 (3.8%), 7 (3.8%) and 8 (17.3%). About 3-4% of poorly resolved material was eluted between HP and HVD. If fraction **A** was warmed on a steam bath for 30-60 min, HPLC revealed only HP (65%), HVD (20.4%) and protoporphyrin (4.3%). Approximately 10% of unknown material was eluted ahead of HP. Also, samples of **A** held at room temperature in water for several days, slowly produced HP detectable by TLC and HPLC in addition to HVD. However most of the material remained intact even after several weeks.

HPLC analysis of commercial hematoporphyrin (HP) revealed HP to constitute 90.4% of the material with HVD comprising 9.3%. The remaining 0.3% was comprised of the same minor components detected in HP although the relative amounts were quite different (**4**, 0.04%; **5**, 0.03%; **6**, 0.015%; **7**, 0.025%; **8**, 0.045%). Using gel filtration columns with various exclusion limits up to 1.5 x 10^6 we found that component **A** was eluted with the same volume as was albumin, indicating **A** to be a polymeric form of the porphyrins, perhaps of a size similar to denatured albumin. The U.V. spectrum of **A** was indistinguishable from that of the mixture except for some broadening of the absorption peaks (Soret band, 370 nm in water).

Hematoporphyrin and protoporphyrin readily aggregate in aqueous solution. These aggregates could be resolved by gel filtration but readily disaggregated upon dilution or under influence of the solvents used for HPLC. This was not true of aggregate **A**. Dilution of this material by a factor of ten or more caused no detectable formation of porphyrin monomers by TLC except for the small amount of HVD found under all conditions. The distribution of components in HPD determined by gel filtration is not changed by using pH buffer (6 to 8) or 0.9% salt solution.

In preliminary testing we have found that **A**, in aqueous solution, is much less fluorescent than the other components of HPD but that the SMT-F tumors in mice given a tumoricidal dose (3.75 to 4.2 mg/kg) are equally fluorescent as those in mice receiving 7.5 mg/kg HPD. We found that the skin photosensitivity induced in albino mice (ICR Swiss) with the tumoricidal dose of **A** (4.2 mg/kg) is much less than that induced by 7.5 mg/kg of HPD (i.e. an equivalent tumoricidal dose). All animals had their rear feet exposed for 30 min to the light source (160 mW/cm) one day following i.p. injection of porphyrin to determine skin sensitivity as previously reported (7,8).

DISCUSSION

The data in Table 1 indicate that essentially all of the tumoricidal activity (as defined here) of HPD resides in minor components **4** through **8** separable by HPLC. None of the major components, HP, HVD or protoporphyrin either isolated from the HPLC columns or tested separately (commercial samples) demonstrated any biologic activity in our system. Protoporphyrin was injected in mice either as an 0.5 mg/ml solution or as a 5.0 mg/ml solution to determine if aggregation state of the porphyrin affected the result. In neither case was a response demonstrated.

It might be expected that minor components **4** through **8** would show more than 2-3 times the activity of HPD, since even taken together they comprise only 20% of the mixture and no biologic activity could be demonstrated in any of the other components. Thus at least a five-fold enhancement might have been expected. HPLC conditions (polar organic solvents) may have degraded some of the active material or not all the components were revealed by HPLC. (In one experiment we found that 85% of the material put onto the HPLC columns was eventually eluted). For this reason and because there had been some indication that the active material may be a dimer or higher porphyrin aggregate, we also carried out gel filtration separations using aqueous media. It could be readily demonstrated that 100% of the material was eluted from these columns. Using a gel filtration column of 20,000 dalton exclusion limit, HPD was readily separated into three major fractions (Table 1). Using columns of larger exclusion limits it appeared that component **A**, in which all the biologic activity resided, was eluted from the columns at approximately the same volume as was albumin, run separately. It appeared that both **A** and albumin were eluted at the void volume of a P-2 or P-10 column (i.e., molecular weight exclusion limit of 2,000 and 20,000 respectively).

The P-10 column appeared optimal for separations. Since **A** and **B** tended to be eluted near each other, HP was occasionally detected as an impurity in **A** (**B** is considered to be mainly aggregates of HP, since it readily dissociated in the TLC system or on HPLC to produce HP). Further, when diluted and rerun on the P-10 column, increasing amounts of monomer (HP) were detectable. On the other hand, **A** did not dissociate even when diluted 10-fold. Also, HP tended to appear in **A** after the fraction remained at room temperature for several days. This may reflect slow dissociation of contaminating HP aggregates or possibly hydration of HVD to produce HP. When freshly collected, HVD was the only material that could be identified in fraction A by TLC. Most of the material remained aggregated and tended to remain at or near the origin of the TLC plates. When dissociated by warming in water, **A** yielded HP, HVD and protoporphyrin with HP predominating. It is not possible to know if HP results from dissociation of HP-containing aggregates or hydration of other porphyrins under these conditions. When subjected to pH < 6, **A** readily produced monomeric porphyrins but was not completely dissociated even after precipitation at pH 3.5 and redissolving in alkali. By gel filtration, at least half the material remains aggregated after these procedures.

It thus appears that the biologic activity of HPD resides mainly or solely in large aggregates of porphyrins at least partially composed of HVD. While HP, HVD and protoporphyrin all tend to aggregate in aqueous media, their aggregates rapidly dissociate upon dilution and in organic solvents. This is not true of component **A**, which could be dissociated only under more stringent conditions, i.e. heating at 100°C in water.

We have also noted that addition of aqueous solution of albumin to aggregates of HP, or protoporphyrin, caused rapid dissociation. (The Soret band shifts from 370 nm to 396 nm upon dissociation in water). However, it required more than twice as much albumin to dissociate a comparable amount of **A**. At this time it is unclear why **A** aggregates are more stable than those composed of HP, HVD or protoporphyrin. The biologic efficacy of **A** may be related to this property, i.e. at least part of the material may reach the tumor interstitial fluid as a large aggregate which may then become associated with the cell membrane and act as a pool of monomeric porphyrin. This would then provide a mechanism for retaining the monomeric porphyrin within the tumor environment. This is consistent with observation of tumor fluorescence following injection of **A** whereas the material in solution is essentially non-fluorescent.

Kessel has recently shown that purified HP is not an effective in vitro sensitizer (6) and that the effects observed with commercial HP can be attributed to HVD and protoporphyrin contaminants. This finding is consistent with our data indicating HP to be inactive in vivo as well. These results of in vitro testing cannot be extrapolated to the in vivo situation since minor hydrophobic contaminants can easily bias the result. In fact, the more hydrophobic materials are not active at all in vivo where ability to localize, as well as elicit a photodynamic process, is required.

The indication of an improved therapeutic ratio is of great significance to reduce or eliminate the induced photosensitivity of patients receiving this material rather than HPD and indicates that the photosensitivity caused by HPD may in fact reside in components other than **A**.

ADDENDUM

It has been suggested that the term hematoporphyrin derivative (HPD) be reserved for the acetylation products of hematoporphyrin, i.e., the mixture produced by the reaction of hematoporphyrin with acetic & sulfuric acids, and that the material generally used *in vivo* be referred to as 'alkali-treated HPD'. This may be appropriate provided that it is realized that hardly anyone uses the acetylated mixture as such, since essentially all attempts to solubilize it result in hydrolysis, whether under acidic, neutral or basic conditions. Thus, when Lipson in his original papers refers to fluorescence of 'HPD' *in vivo*, he really refers to 'hydrolyzed porphyrin acetates', i.e., "alkali-treated HPD' by this terminology. Most investigators have used HPD similarly, including the authors of this paper.

ACKNOWLEDGEMENTS

The authors wish to thank Dr. Kendric Smith for his suggestion of the gel filtration method.

REFERENCES

1. T.J. Dougherty, J.E. Kaufman, A. Goldfarb, K.R. Weishaupt, D.G. Boyle and A. Mittelman, Photoradiation Therapy for the Treatment of Malignant Tumors, *Cancer Res.* *38*:22628 (1978).

2. T.J. Dougherty, G. Lawrence, J.E. Kaufman, D.G. Boyle, K.R. Weishaupt and A. Goldfarb, Photoradiation in the Treatment of Recurrent Breast Carcinoma, *J. Nat'l. Cancer Inst.* *62*:231 (1979).

3. I.J. Forbes, P.A. Cowled, A.S.Y. Leong, A.D. Ward, R.B. Black, A.J. Blake and F.J. Jacka, Phototherapy of Human Tumours using Haematoporphyrin Derivative, Med. J. Aust. 2:489 (1980).

4. Z.P. Pavelic, C.W. Porter, L.M. Allen and E. Mihich, Cell Population Kinetics of Fast- and Slow-growing transplantable Tumors derived from Spontaneous Mammary Tumors of the DBA/2 Ha-DD Mouse, Cancer Res. 38:1533 (1978).

5. A.E. Profio, D.R. Doiron and E.G. King, Laser Fluorescence Bronchoscope for Localization of Occult Lung Tumors, Med. Phys. 6:523 (1979).

6. D. Kessel, Components of Hematoporphyrin Derivative and their Tumor-Localizing Capacity, Cancer Res. 42:1703 (1982).

7. D.J. Borcicky, Repair of Sublethal Damage in Mouse Skin after Photoradiation with Hematoporphyrin Derivative plus Visible Light, M.S. thesis, Roswell Park Memorial Institute, 1976.

8. T.J. Dougherty, D.J. Borcicky, K.R. Weishuapt, A. Goldfarb and R.J. Johnson, Repair of Damage in Experimental and Clinical Photoradiation, Proc., Amer. Assn. Cancer Res. 18:174 (1977).

INTRA-ABDOMINAL APPLICATIONS OF HEMATOPORPHYRIN PHOTORADIATION THERAPY

H.O. Douglass Jr., H.R. Nava,
K.R. Weishaupt, D. Boyle, M.G. Sugerman,
E. Halpern and T.J. Dougherty

Roswell Park Memorial Institute
Buffalo, New York

INTRODUCTION

At Roswell Park Memorial Institute (RPMI), long experience in the management of tumors with hematoporphyrin derivative (HPD) photoradiation therapy (PRT) has documented the destruction of a wide range of neoplasms with remarkable regularity (1,2). The surrounding inflammatory response and eschar that formed at the site of the tumor nodule indicated that some damage to normal adjacent tissue occurred. In an attempt to evaluate the potential for HPD PRT in the abdomen, we carried out preclinical trial of the tolerance of normal intra-abdominal viscera, and of that containing an implanted Brown-Pearce tumor, in rabbits. For an assay in a primate system, three baboons were studied. Lessons learned from animal trials have recently been applied toward the management of intra-abdominal human neoplasia.

ANIMAL STUDIES

Photoradiation therapy studies were performed in 12 young (1.4-1.8 kg) male belted Dutch rabbits (including one exposed to 631 nm red light but who did not receive HPD, and one who received HPD but no light.) Of the ten rabbits who received both HPD and PRT, the interval between drug administration and PRT was varied in 24 hr increments between 48-96 hr (one at 48 hr, six at 72 hr and four at 96 hr).

Since most of the clinical studies of superficial lesions were treated with PRT 72 hr after HPD, the predominant proportion of these rabbit studies were carried out in the 72 hr rabbit group. HPD was administered intravenously at 5 mg/kg body weight.

Freshly harvested Brown-Pearce tumor, carried in the anterior chamber of the eyes of rabbits, was implanted by trocar into the submucosa of bowel, liver, pancreas or bladder of half of the treatment sites in each rabbit, and allowed to grow 7 mm in diameter as confirmed by laparotomy under anesthesia prior to HPD injection.

At time of PRT, the rabbits were anesthetized and their incisions reopened. Illumination of the peritoneal cavity with a Wood's dermatologic ultraviolet light and examination through goggles equipped with an amber filter generally showed fluorescence of the tumor implants only, with a "halo-like" pattern at the tumor borders. PRT was administered using a tunable dye laser with light delivered through 200 micron diameter quartz glass fibers to the tumor site, illuminating a one-centimeter diameter site on the serosal surface including the tumor or an equivalent area of normal tissue, with red light of 625-640 nm (peak at 631 nm), the fiber supported by a stationary stand. Light intensity was measured before and after each treatment with Coherent model 210 light meter. When PRT exposure exceeded 10 min., the tissues were moistened with saline to prevent tissue damage from drying. Light intensities utilized were 260, 300, 400, 500, 550, 600, 750, and 1400 mW/cm^2) administered over 2-20 min. Lower light intensities (260-600 mW/cm^2) were usually utilized in association with shorter treatment intervals; with higher intensities (400-1400 mW/cm^2) utilized in animals that had received HPD 96 hr prior to PRT.

Three baboons were each treated in one gastric and one colonic site following injection of 2.5 or 5.0 mg of HPD 72-96 hr earlier. In the primate, the bowel was opened and treatment administered intralumenally. Exposures in the baboon ranged from 120-400 mW/cm^2 for 2.5-11 min.

All animals were explored 5-7 days following PRT for excisional biopsy and histologic examination of the treated areas. Damage to normal tissue was graded (Table 1); Grade III and IV damage is considered unacceptable, while Grades I and II represent tolerable levels of tissue damage.

RESULTS

Immediately following PRT, only tumors treated with

Table 1. Histological grading system

Grade	Damage	Histology
IV	Complete	-full thickness necrosis -loss of mucosa and destruction of smooth muscle
III	Subtotal	-coagulation necrosis and hemorrhage -outline of muscularis remaining -viable cells remaining -granulation tissue
II	Slight	-mucosa cells not nucleated -architecture of muscularis intact
I	None	-no damage -no necrosis

more than 500 mW/cm² for more than 10 min. showed any gross change: a discoloration to either lighter, almost white, or darker, almost black from the normally reddish appearance of healthy tumor. No discoloration or other changes were seen in normal tissue. At time of excision, no further gross changes were apparent (all treated areas had been marked with a fine suture at the margin of the illuminated circle for later identification), except apparent necrosis of tumor.

In the rabbit, a light dose of >40 milliwatt-hr per square centimeter (144 J/cm²) was generally associated with an unacceptable level of normal tissue destruction; levels above 137 mW-hr uniformly resulted in tissue death (grade IV). PRT doses of 40 mW-hr or less generally resulted in minimal damage to normal tissues. Normal stomach, small bowel, colon and pancreas in the rabbit and baboon all had similar resistances to PRT, while normal liver, kidney and bladder were found to be more sensitive. Normal liver appeared to be the most sensitive organ, with hepatocytes preserved only in the immediate vicinity of peripheral portal triads (Table 2).

In the baboon, PRT doses of 37.5 mW-hr or higher resulted in perforation of the colon, while at a dose of 10 mW-hr, only submucosal edema was seen. In one baboon, 47 mW-hr caused colonic hemorrhage, this same dose causing no histologic change in shaven skin. The normal baboon stomach appeared slightly more resistant to PRT damage than the colon, whereas in the rabbit, the opposite was true. Marked hepatocyte damage was noted in the baboon at a PRT dose as low as 12 mW-hr.

The conclusions that can be drawn from these preclinical studies are that gastrointestinal PRT should take place 72 or more hr following HPD administration, that the dose should be limited to 40 mW-hr or less at 72 hr, but can be increased to 80 mW-hr at 96 hr. Although the liver is quite sensitive to HPD-PRT, administration of no more than 25 mW-hr at least 72 hr following HPD may yield a tolerable level of damage to the normal liver. Hepatocytes in the area of larger portal triads are least sensitive to HPD-PRT.

HUMAN STUDIES

To date, four patients with gastric cancers, two with colorectal neoplasia and one esophageal cancer have been treated with HPD-PRT (Table 3), administered via the same fiberoptic technique as utilized in the animals but with 300 mW being provided through each of three implanted fibers, or a single transendoscopic fiber which delivered 100 mW/cm^2, one centimeter from the probe, the exposed surface being 3 cm in length (with 360° diffusion of light).

Patients were treated with 5 mg/kg of drug 72 hr prior to PRT. Fibers were placed or implanted to provide overlapping coverage of the tumor. The dose in any site averaged 30 mW-hr although in areas of overlap, a dose as high as 40 mW-hr may have been administered.

From the results of these preliminary human trials, variable tumor sensitivity is apparent. To prevent tumor excavation with resulting perforation of adjacent vital organs, it may be desirable in selected cases to reduce the dose of HPD to 2.5 mg/kg body weight or increase the interval between HPD and PRT to 96 hr.

DISCUSSION

Although these preliminary studies have suggested that

Table 2. Results of PRT in rabbits

ORGAN	Interval between HPD & PRT (hr)	Extent of Normal Tissue Damage	
		Grade I-II	Grade III-IV
Stomach	48		1 (42)
	72	2* (40+, 75)	3 (40,138,150)
	96	1 (83)	2 (125,133)
Small Intestine	48	3 (24,24,38)	
	72	5 (10,25,25,40,75)	5 (20,40,125,125,150)
	96	1 (63)	1 (125,150,150)
Liver	48	1 (26)	1 (24)
	72	2 (20,25)	2 (138,150)
Pancreas	96	1 (125)	
Kidney	72		1 (40)
Bladder	72		2 (80,150)

* number of animals

\+ PRT in mW-hr/cm²; (40 mW-hr/cm²=144 Joules/cm²).

Table 3. Initial human studies

Tumor Type	Tumor Site	Histology	Treatment Objective	Treatment Method	Result
1. gastric	subcutaneous	adenoca.	efficacy	implanted fibers	>90% slough (necrotic eschar)
2. colon	right abdomen recurrence	adenoca.	adjuvant to palliative excision	implanted fibers - repeat via percutaneous catheters	tumor recurred; fecal fistula (late)
3. rectum	pelvic mass, recurrence	adenoca.	adjuvant to palliative resection	implanted fibers in residual lymph nodes	necrosis in one node, no damage in another
4. stomach	esophageal recurrence	adenoca.	palliation -permit deglutition	transendoscopic fibers (3 sites)	massive tumor slough, tracheo-esophageal fistula
5. esophagus	mid esophagus	squamous	preoperative	transendoscopic fiber	pain, fever; little histologic change at 48 hr (excised)
6. stomach	cardio-esophageal junction	adenoca.	preoperative	transendoscopic fiber	necrosis in depth of tumor, excised at 72 hr
7. stomach	esophageal recurrence	adenoca.	preoperative	transendoscopic fiber	not resectable, tumor slough, improved deglutition. Broncho-esophageal fistula after 3 weeks

a practical intra-abdominal application of HPD-PRT may be developed, further pilot studies will be necessary to enhance the safety of the approach. It may be necessary to evaluate each tumor and each site separately.

A major question that needs to be approached relates to the problem of the sensitivity of malignant vs normal liver cells to HPD-PRT. If it can be shown that tumors are more sensitive than human liver, a much needed therapeutic approach to the treatment of liver metastases could be developed.

Another exciting area for evaluation is that of the pancreas. Although only one rabbit and one baboon pancreas were studied, the suggestion that the normal pancreas is quite resistant to HPD-PRT suggests the possibility of treating patients with pancreatic neoplasia.

Intra-abdominal cancers might at first seem poor candidates for HPD-PRT. The advent of ultrasonically guided needles may obviate this problem if efficacy of PRT in intra-abdominal cancer can be demonstrated. It is possible that PRT may in the future become one more tool in the oncologists armament of multi-disciplinary therapy.

REFERENCES

1. Dougherty, T.J., Kaufman, J.E., Goldfarb, A., Weishaupt, K.R., Boyle, D., and Mittelman, A. Photoradiation Therapy for the Treatment of Malignant Tumor. Cancer Res. 38:2628-2635, 1978.

2. Dougherty, T.J., Gomer, C.J., and Weishaupt, K.R. Energentics and Efficiency of Photoinactivation of Murine Tumor Cells Containing Hematoporphyrin. Cancer Res. 36:2330-2333, 1976.

EFFECTIVENESS OF HPD AND RADIATION THERAPY IN LUNG CANCER

Harubumi Kato, Chimori Konaka, Jutaro Ono, Yasushi Matsushima, Katsuaki Nishimiya, Johnson Lay, Hiroyuki Sawa, Hideki Shinohara, Tsutomu Saito, Komei Kinoshita, Takahisa Tomono, Motohiko Aida and Yoshihiro Hayata

Department of Surgery
Tokyo Medical College
6-7-1 Nishishinjuku, Shinjuku-ku
Tokyo 160, Japan

INTRODUCTION

The diagnostic rate of lung cancer has been improved by the development of the fiberoptic bronchoscope and this instrument has particularly promoted the detection of early stage of central type lung cancer. But therapeutic results have been unsatisfactory, partly because of difficulties in resecting cases with poor pulmonary function. On the other hand, chemotherapy and radiation therapy can have serious side effects. New therapeutic modalities are therefore required for treatment of early stage inoperable lung cancer cases.

Photoradiation therapy (PRT) for malignant tumors has been pioneered by Dougherty and his coworkers (1,2). This therapy is based on the activation of hematoporphyrin derivative (HPD) by light. It is well known that HPD has an affinity for malignant tissue and can mediate light-catalyzed cytocidal effects.

We have investigated PRT using laser equipment as a light source and intravenous HPD administration in experimentally-induced canine lung cancer and recognized its ther-

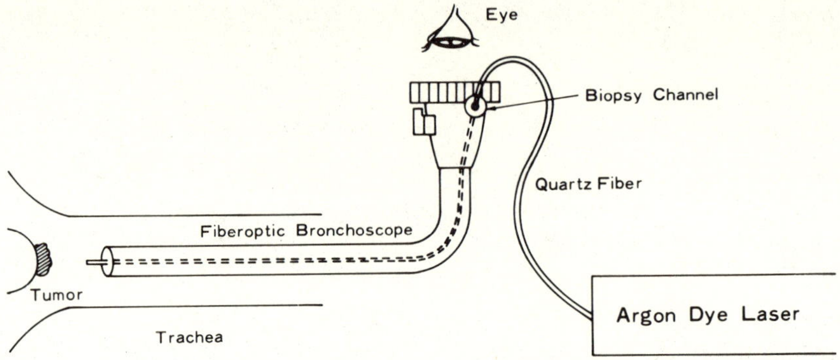

Figure 1. Schema of therapeutic procedure. An argon dye laser beam is transmitted via a 400 micron quartz fiber inserted through the instrumentation channel of the fiberoptic bronchoscope. The cancer lesion is irradiated at a distance of 1-5 cm from the fiber tip with an average laser power of 200 mW for 20 min.

apeutic effectiveness (3). We subsequently employed this new technique for therapy of human lung cancer cases.

MATERIALS AND METHODS

Hematoporphyrin derivative

HPD was provided by Dr. T.J. Dougherty, Department of Radiation Biology, Roswell Park Memorial Institute, Buffalo, New York, who prepared it by means of a modification of Lipson's method (4). The half-life of HPD in blood is approximately 25-35 hr for a dose of 2.5-4.0 mg/kg body weight (5). In our studies, 2.0-5.0 mg/kg of HPD was injected intravenously 48-72 hr prior to PRT.

Therapeutic equipment and procedure

The light source was an argon laser model 171-08, 15W, 457.9-514.9 nm (Spectra Physics Co., Mountain View, CA) coupled to a dye laser (model 373-03) using Rhodamine B dye. The 514 nm argon laser beam was converted to a near 630 nm beam by the dye laser. The quartz fiber (400 micron, Quartz Products Co. Plainfield, N.J.) was coupled to the laser using a fiber interface. The laser beam was transmitted (Figs. 1, 2) via the quartz fiber inserted through the instrumentation channel of an Olympus BF-1T fiberoptic bronchoscope (6-8).

Irradiation time varied from 5-30 min. with a power of

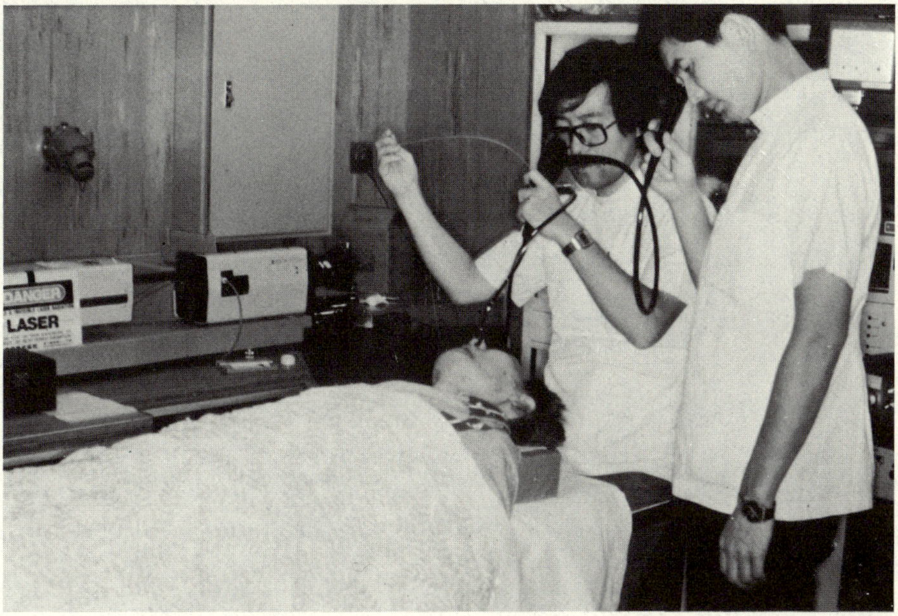

Figure 2. The photoradiation therapy is performed following local anesthesia by 4% xylocaine spray. The case is 59-year-old female with early stage central type lung cancer at the right upper lobe bronchus.

90 to 800 mW at the fiber tip. From 1-6 irradiations were performed 48-336 hr after i.v. administration of 2.0-5.0 mg/kg of HPD. Photoradiation was performed from a distance of 1-5 cm in cases in which tumor had developed invasively replacing the bronchial mucosa. Since the target moves with respiration, the normal epithelium is also irradiated, but this does not present a problem due to the extremely low power of the beam. Where tumor had developed in the bronchial lumen, the quartz fiber was implanted into the tumor. The angle of the irradiation beam can be varied from 22.6-53.0 degrees by a fiber-pressing and bending mechanism, and can be adjusted during the procedure, depending on the target area size. Photoradiation therapy was performed with the same local anesthesia (4% Xylocaine spray) as in conventional fiberoptic bronchoscopic examinations (6-8).

Clinical cases

Forty-one cases of lung cancer and 2 cases of severely atypical squamous metaplasia were treated; 38 were male and 5 were female. The age distribution was 38-79. According

Table 1. Use of laser photoradiation in lung cancer

	Epidermoid ca.	Adeno ca.	Large ca.	Small ca.	Other ca.	Total
Early Stage	3	0	0	0	0	3
Stage I	4	0	0	0	0	4
Stage II	6	1	0	0	0	7
Stage III	12	2	2	1	0	17
Stage IV	4	2	2	1	1	10
Total	29	5	4	2	1	41

Severely Atypical Squamous Metaplasia 2

to clinical stage, 3 were early, 4 were stage I, 7 were stage II, 17 were stage III and 10 were stage IV.

Histologically, there were 29 cases of squamous cell carcinoma, 5 cases of adenocarcinoma, 4 cases of large cell carcinoma, 2 cases of small cell carcinoma and 1 case of bronchial metastasis of malignant melanoma, (Table 1). PRT was performed preoperatively in 14 cases to examine histopathological effects. In 9 of these cases, PRT was performed to reduce the extent of resection preoperatively. All early stage lung cancer cases showed no abnormal chest X-ray film findings and were diagnosed by sputum cytology. All were squamous cell carcinomas; fiberoptic bronchoscopy was subsequently performed for localization of tumor site.

The first case of early stage lung cancer was a 74-year-old male who had complained of cough and sputum for 3 months. His chest X-ray film showed no abnormalities but his sputum cytology showed well differentiated squamous cell carcinoma. Early stage cancer was suspected on the basis of cell features such as the cells being small and round. Fiberoptic bronchoscopy showed two tiny tumors, sized 2x2 mm, one in right B_2b and one at the bifurcation between B_2a and B_2b bronchi. Cytological brushing and histological biopsy yielded a definitive diagnosis of well differentiated squamous cell carcinoma. He adamantly refused surgery.

The second case was a 76-year-old male who also presented with cough and was X-ray negative, sputum cytology positive. A superficially irregular tumor was detected at the bifurcation of left $BP_{10}ai$ and ii bronchi by fiberoptic bronchoscopy. Cytological and histological findings showed well differentiated squamous cell carcinoma. His pulmonary function was very poor due to pulmonary emphysema: VC:1.35l, %VC:44.2%, %FEV_1:46.5%.

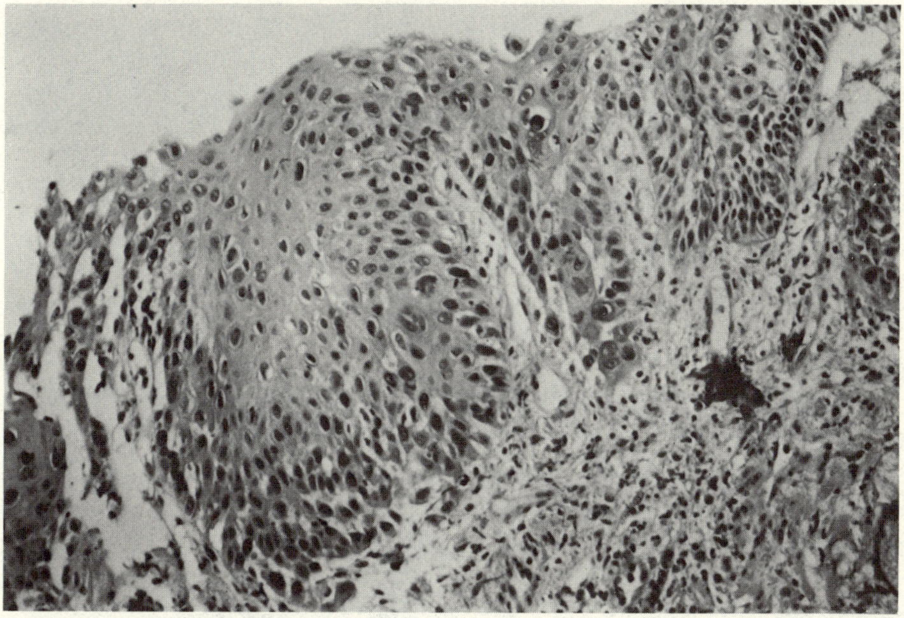

Figure 3. This specimen was obtained by endoscopic biopsy and reveals well differentiated squamous cell carcinoma without submucosal invasion. This early stage case was detected by sputum cytology and showed no abnormalities in the chest X-ray film.

The third case was a 61-year-old female with sputum and cough for 6 months. Squamous cell carcinoma cells were detected in her sputum but her chest X-ray showed no abnormalities. A tumor was observed at the bifurcation of the segmental bronchi of the right upper lobe. Endoscopy revealed thickened mucosa; brushing cytology and biopsy showed well-differentiated squamous cell carcinoma (Fig. 3). Surgery was not indicated because of her poor pulmonary function. Her function test showed VC:1.531, %VC:70.1%, $\%FEV_1$:58.2%

One case of severely atypical squamous metaplasia was a 61-year-old female heavy smoker. An infiltrative shadow was found in the right upper lung field. Fiberoptic bronchoscopy showed slight redness and thickening at the right upper lobe bronchus. Brushing cytology showed findings suggestive of malignancy: frequent mitosis, hyperchromasia, coarse chromatin and irregular nuclear size. A histologic study showed atypical squamous metaplasia (Fig. 4).

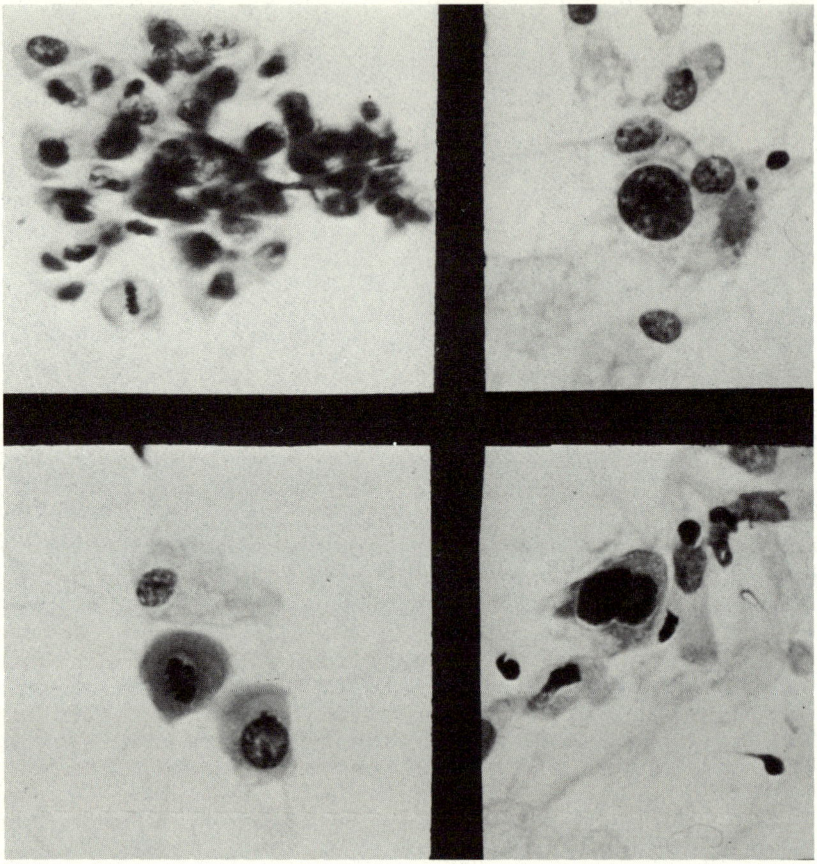

Figure 4. This brushing cytological specimen was obtained from a 61-year-old female who presented with cough. Fiberoptic bronchoscopy showed slight redness and thickening of the mucosa of the right upper lobe bronchus. Cell features reveal frequent mitosis, hyperchromasia, irregular nuclear size. Biopsy revealed atypical squamous metaplasia. This case was treated with 150 mW of argon dye laser beam for 15 minutes. The atypia disappeared completely after 2 weeks of therapy.

The other case of severely atypical squamous metaplasia was a 46-year-old male who had complained of sputum. Severely atypical squamous metaplasia was detected by sputum cytology. Fiberoptic bronchoscopy showed mucosal irregularity and thickening were observed at the right main bronchus.

Table 2. Results of PRT by clinical stage

	Case	Resection	Complete Remission	Partial Remission
Early Stage	3	0	3	0
Stage I	4	2	1	3(2)*
Stage II	8	6	0	8(2)*
Stage III	16	6	1	15
Stage IV	10	0	0	10
Total	41	14	5	36(4)*
Severely Atypical Squamous Metaplasia	2	0	2	0
Total	43	14	7	36

( )* : Significant Remission

This case showed congenital anomalous bifurcation of the right upper lobe bronchus immediately after the carina. Brushing cytology and biopsy revealed severely atypical squamous metaplasia.

Evaluation of results

The therapeutic effects were evaluated by histology, brushing cytology and endoscopy in cases not undergoing surgery. More detailed examinations were performed histopathologically in surgically-resected cases and autopsied cases. Effectiveness was divided into the following 4 categories. Complete remission: no evidence of tumor by endoscopy, brushing cytology or histology. Significant remission: 60-<100% of the original tumor volume disappeared endoscopically. Partial remission: <60% of the original volume disappeared endoscopically. No remission: no endoscopically recognizable change of tumor (7,8).

RESULTS

The therapeutic effects on 52 foci in 41 cases examined by the above criteria. Complete remission was obtained 12 foci (23.1%), significant remission in 24 foci (46.1%), and partial remission in 16 foci (30.8%). Opening of bronchi that had been obstructed by tumor was obtained in 11 out of 23 foci (47.8%). The therapeutic results according to clinical stage are shown in Table 2.

All early stage cases showed complete remission (Fig. 5). These cases are disease free at 22, 11 and 10 months, respectively. They were treated by photoradiation only. One stage I case showed complete remission endoscopically but a tumor shadow 1 cm in diameter remained roentgenologi-

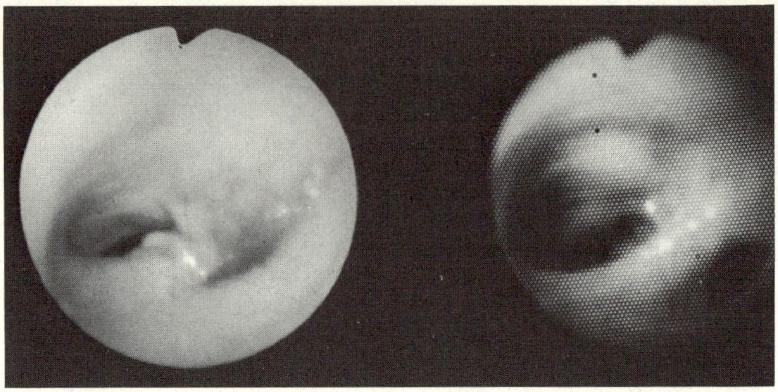

Figure 5. This case is a 76-year-old male who suffered from sputum and cough. His chest X-ray film showed no abnormalities but sputum cytology revealed squamous cell carcinoma. Fiberoptic bronchoscopy showed a slight irregular superficial tumor at the bifurcation between the left B_{10}ai and ii bronchi as shown in this left figure. The right figure shows the findings 3 months after photoradiation therapy and no abnormalities can be recognized. He is disease free at 11 months.

cally in the left S_3 area. Therefore 30 Grays radiation by Linac was performed following which the tumor shadow disappeared. This case was inoperable due to poor pulmonary function and is disease free at 8 months. The single stage III case which showed complete remission was a case of squamous cell carcinoma of the trachea. No tumor shadow was observed on the chest X-ray film but Linac irradiation was performed 2 months after PRT on the suspicion of extra tracheal invasion. The 2 cases of severely atypical squamous metaplasia showed complete disappearance of atypia. Table 3 indicates the therapeutic effectiveness according to histologic type. Complete remission was obtained in 9 squamous cell carcinoma foci (26.5%) and in 3 adenocarcinoma foci (27.3%). Significant remission was obtained in 16 squamous cell carcinoma foci (47%), 5 adenocarcinoma foci (45.5%) and 3 large cell carcinoma foci (75%). There were no cases of complete or significant remission in small cell carcinoma or the single case of malignant melanoma. Table 4 shows the effectiveness of PRT to reduce the extent of resection area and to make inoperable cases operable. PRT was performed preoperatively in 9 cases of advanced lung cancer. This purpose was successively achieved in 6 cases. Three out of 4 inoperable cases could be resected, with sleeve lobectomy performed in 2 cases and pneumonectomy in one case. Three lobectomies were performed in 5 cases in which pneumonectomy was indicated prior to photoradiation.

Table 3. PRT results by according to histological type

	Sq. Ca.	Adenoca.	Large Ca.	Small Ca.	Malig. Mela.	Total
C.R.	9	3	0	0	0	12
S.R.	16	5	3	0	0	24
P.R.	9	3	1	2	1	16
N.R.	0	0	0	0	0	0
Total	34	11	4	2	1	52

C.R.: complete remission, S.R.: significant remission,
P.R.: partial remission, N.R.: no remission

Table 4. Results of PRT in combination with surgery.

Case	Age	Location	Conventional Approach	PRT+Surgery
K.I.	54	r.u.l.br. -trachea	(inop.)	sleeve resection at r. main br.
M.K.	65	l.l.l.br. -l.main br.	(l. pneumonectomy)	l. lower lobectomy
H.I.	60	r.u.l.br. -r.main br.	(inop.)	sleeve resection at r. main br.
J.T.	64	l.main br. -carina	(inop.)	l. pneumonectomy
F.K.	64	r.B3-r.main br.	(r. pneumonectomy)	r. pneumonectomy
T.O.	63	r.main br.	(inop.)	probe thoracotomy
T.S.	62	r.main br.	(r. pneumonectomy)	r. pneumonectomy
Y.N.	51	l.l.l.br. -l.main br.	(l.pneumonectomy)	l. lower lobectomy
T.Y.	54	l.l.l.br. -l.main br.	(l. pneumonectomy)	l. lower lobectomy

In a 64 year-old male who complained of cough and sputum, the primary tumor, sized at 5x3.5 cm, was located in the left S_3 area on chest X-ray film. Superficial tumor invasion was observed in the left main bronchus up to 1 cm from the carina. Since surgery is only indicated if invasion is no closer than 2 cm from carina, this case was inoperable. The histologic type was well differentiated squamous cell carcinoma. PRT was performed using a 200 mW beam for 30 minutes to the left main bronchus. The tumor invasion had disappeared 2 weeks later and pneumonectomy could be performed a month after PRT (Fig. 6). No tumor cells were observed microscopically in the bronchial stump.

In another case, a 64-year-old female complained of

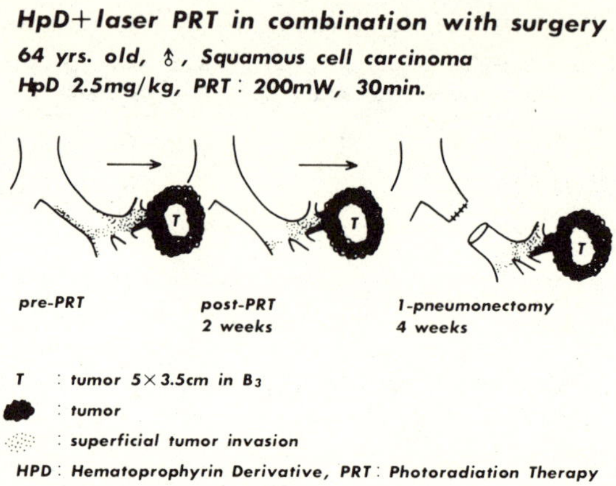

Figure 6. This case is a 64-year-old male. Invasion 1 cm from the carina regressed after phototherapy and it was then possible to perform a left pneumonectomy.

dyspnea, cough and sputum. Her chest X-ray film showed atelectasis of the right middle and lower lobes. Bronchography showed obstruction of the right main bronchus (Fig. 7), and fiberoptic bronchoscopy showed a large polypoid adenocarcinoma tumor occupying the entire lumen of the right main bronchus. Intratumoral irradiation (9) was performed using a 600 mW beam for 30 min., 72 hr after i.v. administration of HPD. Significant decrease of the tumor was recognized with opening of the right main bronchus the day following photoradiation and her clinical symptoms improved. Complete disappearance from the main bronchus was obtained a week after this therapy but the original primary tumor remained in distal B_3 (Fig. 8). The chest X-ray film taken 14 days after the procedure showed the disappearance of atelectasis and the tumor shadow could be recognized in the S_3 area (Fig. 9). Upper lobectomy was planned but the wide expansion of the tumor in the hilum necessitated pneumonectomy on thoracotomy. Detailed histological examination of the resected specimen was performed. The tumor was histologically well-differentiated adenocarcinoma without any invasion of the proximal bronchial wall.

A 62-year-old male whose chest X-ray showed a tumor shadow in the right hilum and atelectasis of the middle lobe bronchus was treated with this procedure at the obstructed area of the middle lobe bronchus using a 180 mW beam for 30 min. Opening of the middle lobe bronchus was obtained and surgery was performed 1 week after the procedure. The surgically resected specimen revealed that the intrabronchial tumor was almost completely necrotized, but the extrabron-

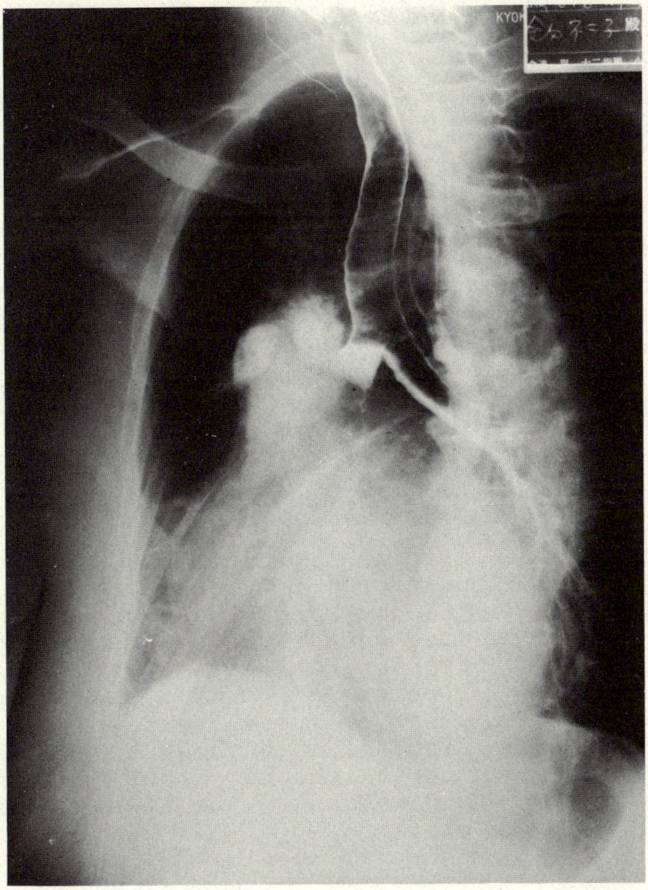

Figure 7. A 67-year-old female complained of dyspnea. Bronchography showed complete obstruction of the right main bronchus and atelectasis of the middle and lower lobes.

chial tumor showed no effects of PRT (Fig. 10).

Histopathological changes recognized after treatment included degeneration of cancer cells: nuclear and cytoplasmic changes such as nuclear vacuolization, fragmentation of nuclear chromatin, homogenization of chromatin, cytoplasmic keratinization and vacuolization. Infiltration by inflammatory cells, hemorrhage and edema were observed in the necrotic area and surrounding connective tissue.

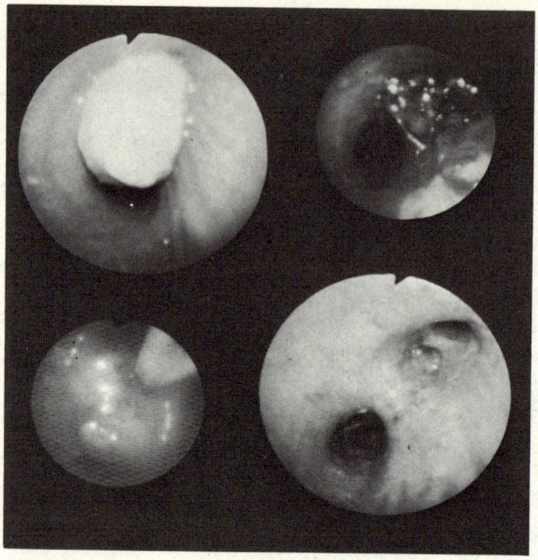

Figure 8. The upper left photograph shows a large polypoid tumor occupying almost the entire right main bronchus. Histologic studies indicated a well-differentiated adenocarcinoma. One day after intratumor irradiation (left lower), a remarkable decrease of tumor and opening the the main bronchus were observed (upper right). The lower right shows the findings 1 month after PRT. We had planned an upper lobectomy, but the wide expansion of tumor involvement in the hilum necessitated pneumonectomy.

DISCUSSION

Methodology and results of PRT are described in this paper. The argon dye laser used in these clinical trials provides optimal wavelength for therapy (near 630 nm) since this has optimal light penetration characteristics for malignant tissue. Shorter wavelength light also excites HPD but is absorbed by blood and melanin, thereby limiting penetration. To treat malignant tumors, penetrating light is required because tumor tissue often has more vascularization and frequently shows hemorrhage on the surface. Also the depth of tumor invasion cannot be determined on the basis of endoscopy findings alone. It was shown that the cytocidal mechanism is due to a photodynamic reaction involving oxidation of tissues via porphyrin-catalyzed production of singlet oxygen (11).

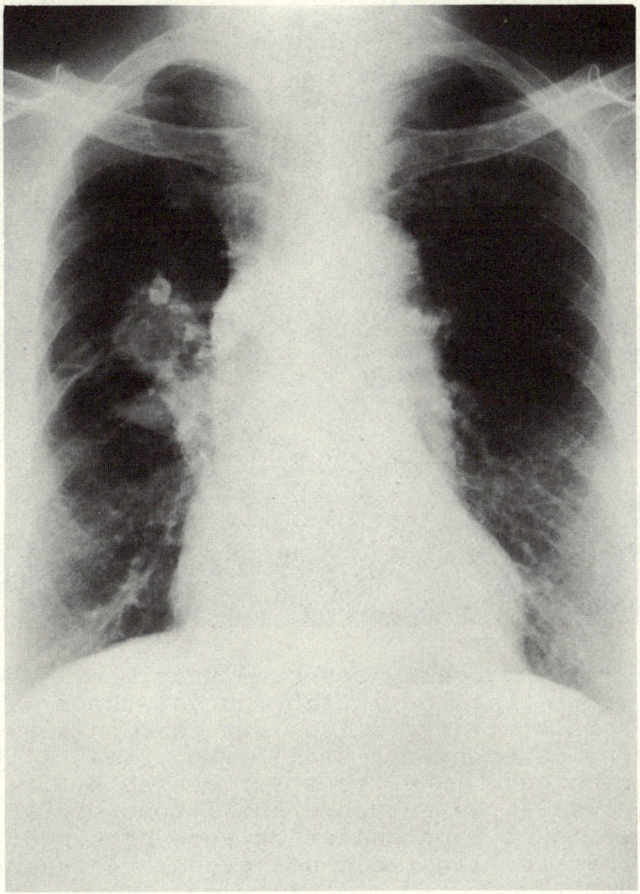

Figure 9. This chest X-ray 1 month after PRT showed that the aletectasis had completely disappeared and the primary tumor could be recognized in the middle lung field.

Our results showed that responses were obtained in all cases. In particular, 3 cases of early stage lung cancer showed complete remissions and no tumor recurrence has been observed for 22 months, 11 months and 10 months respectively. This means that it is possible to curatively treat early-stage lung cancer by this method alone.

Lung cancer incidence is much higher in older persons who frequently have poor pulmonary or cardiac function. In such inoperable cases, other therapies such as radiotherapy and chemotherapy, can be harmful due to the side effects.

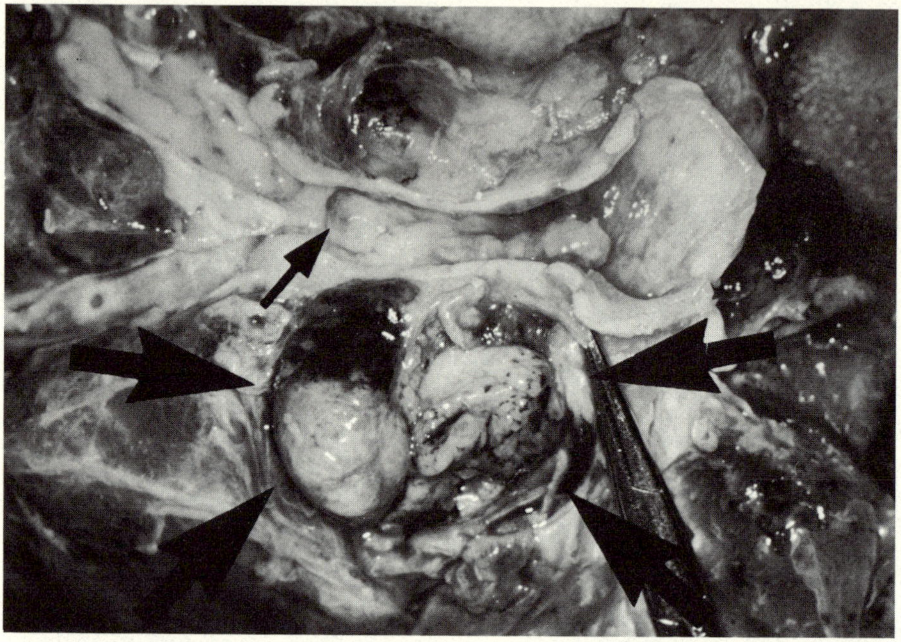

Figure 10. The surgically resected specimen 1 week after PRT shows that the intrabronchial tumor was almost completely necrotized but that in a small area, indicated by the small arrow, viable tumor remained. The focus surrounded by the large arrows is the extrabronchial tumor in which no effects of photoradiation were obtained. This means that the light did not penetrate the bronchial wall.

From that point of view, this new therapy is significantly valuable, because it has no serious side effects. PRT can also be used to reduce the extent of resection by preoperative photoradiation. It was effective when used in this way in 6 cases out of 9 cases. Therefore this method may be effective in increasing 5 year survival. PRT is also effective in obtaining local effects in advanced stage cancer cases, thereby improving the performance status.

In terms of complications, slight sunburn was recognized in 90%, the degree depended upon the dosage of HPD. A case that received 2.5 mg/kg body weight HPD intravenously went out under the sunshine for 3 hr after injection of HPD, despite being enjoined not to. He complained of facial pain and acute sunburn; redness and grade I edema were observed but the changes disappeared after 2 days. Two cases of obstructive pneumonia were observed after PRT due to obstruction of the normal bronchus by highly viscous exudate. This substance looked like a grayish-white gelatinous polipoid growth, exuded from necrotic tissues and could be observed obstructing bronchi a few days after PRT. Fiberoptic bronchoscopy should be performed every day for a few days after the PRT session so as to remove these secretions to avoid obstructive pneumonia. Two cases of bronchial fistula were recognized after intratumoral photoradiation in cases of bronchi obstructed by tumor invasion.

Examining results with regard to histology showed that better effectiveness was observed in squamous cell carcinoma and adenocarcinoma than in other histologic types. The effect might be related to the more abundant cytoplasm in squamous cell carcinoma and adenocarcinoma, since HPD may adhere to the cytoplasmic membrane or localize in the cytoplasm.

Indications for this method should be limited to the treatment of early stage lung cancer cases, especially in inoperable cases, with tumor in a location that can be irradiated by a quartz fiber inserted through the instrumentation channel of the fiberoptic bronchoscope. This method, as preoperative treatment, can also be indicated in cases of superficial mucosal invasion, to reduce the resection area (except in cases of invasion of the entire bronchial wall). Finally, this procedure can be indicated in advanced cases to improve performance status. It is also possible to treat severely atypical squamous metaplasia.

Photoradiation therapy should not be performed in cases in which invasion to the pulmonary artery is suspected. In these cases, photoradiation would be dangerous. There are still problems concerning the illumination field angle, obtaining better light penetration and decreasing patient light sensitivity. One of the most attractive advantages of this therapy is that it enables selective treatment of malignant tissue without any damage to normal tissues. Our results clearly demonstrate the effectiveness of this modality in the curative treatment of early stage cases, in increasing the effectiveness of surgery and in local therapy to improve performance status.

SUMMARY

The authors have performed HPD and laser photoradiation therapy (PRT) in 41 cases with lung cancer and 2 cases of severely atypical squamous metaplasia. HPD (2.0-5.0 mg/kg) was administered intravenously 48-72 hr prior to PRT therapy. An argon dye laser was used as light source for excitation. The laser beam was transmitted via a 400 micron quartz fiber inserted through the instrumentation channel of a fiberoptic bronchoscope. The lesions were irradiated superficially (90-600 mW, 10-30 min), or intratumorally (200-800 mW, 5-15 min).

Three patients were in early clinical stage, 4 were stage I, 7 were stage II, 17 were stage III and 10 were stage IV. The histologic type was squamous cell carcinoma in 29, adenocarcinoma in 5, large cell carcinoma in 4, small cell carcinoma in 2 and a single case of bronchial metastasis of malignant melanoma.

Therapeutic effectiveness was evaluated as follows; Complete remission (CR): no evidence of tumor endoscopically, cytologically and histologically. Significant remission (SR): more than 60% reduction of the original tumor volume but less than 100%. Partial remission (PR): <60% reduction of the original tumor volume. No remission (NR).

CR was obtained in all cases of early stage lung cancer and in one case each for stage I and stage III. These cases remain disease free at 22 months, 11 months and 10 months, respectively. Two cases of severely atypical squamous metaplasia volunteering for treatment, showed complete disappearance of atypia.

The effectiveness of the PRT was also evaluated in combination with surgery. PRT was performed preoperatively in advanced lung cancer cases to reduce the extent to be resected and to make inoperable cases operable. Effectiveness was observed in 6 out of 9 cases. It was recognized that one of the important uses of this modality is to increase the indications for surgery and minimize the degree of aggressiveness necessary. There remain problems concerning the radiation angle, obtaining better light penetration and decreasing skin photosensitization, but this therapeutic modality has shown its effectiveness.

ACKNOWLEDGEMENTS

The authors wish to express their deep gratitude to Dr. Thomas J. Dougherty and his staff at the Roswell Park Memorial Institute, Kenneth R. Weishaupt of Oncology Research and Development, Inc. who supplied us with HPD and provided much helpful advice and to Associate Prof. J.P. Barron of St. Marianna Univ. School of Medicine for his review of the manuscript.

REFERENCES

1. Dougherty TJ, Kaufman JH, Goldfarb A, Weishaupt RK,

Boyle D, and Mittleman A: Photoradiation Therapy for the Treatment of Malignant Tumors. Cancer Res., 38:2628-2635, 1978.

2. Dougherty TJ, Lawrence G, Kaufman JH, Boyle D, Weishaupt KR, and Goldfarb A: Photoradiation in the Treatment of Recurrent Breast Carcinoma. J. Natl. Cancer Inst., 62: 231-237, 1979.

3. Hayata Y, Kato H, Aida M, Konaka C, Ono J, Saito T, Iimura I, Matsushima Y, Yoneyama K, and Nishimiya K: Hematoporphyrin derivative and photoirradiation therapy with argon-dye laser for the treatment of experimentally induced central type lung cancers in dogs. Igakunoayumi (in Japanese) 116:203-205, 1981.

4. Lipson RL and Baldes J: Photodynamic Properties of a Particular Hematoporphyrin Derivative. Arch. Dermatol. 82-508-516, 1960.

5. Gomer CJ and Dougherty TJ: Determination of [^{3}H]- and [^{14}C]-Hematoporphyrin Derivative Distribution in Malignant and Normal Tissue. Cancer Res. 39:146-151, 1979.

6. Hayata Y, Kato H, Konaka C, Aida M, Iimura I, and Tahara M: Hematoporphyrin derivative and laser photoradiation system in the diagnosis and treatment of lung cancer. Proc. 7th Asia-Pacific Congr. Dis. Chest. 33, 1981.

7. Hayata Y, Kato H, Konaka C, Ono J, and Takizawa N: Hematoporphyrin derivative and laser photoradiation system in treatment of lung cancer. Chest (in press).

8. Kato H, Konaka C, Ono J, Matsushima Y, Nishimiya K, Hakoshima A, and Hayata Y: Laser photoradiation with hematoporphyrin derivative in lung cancer. Proc. 4th Congr. Int. Soc. Laser Surgery, Laser-Tokyo '81, 14-5 -14-8, 1981.

9. Dougherty TJ, Thoma RE, Boyle DG, and Weishaupt KR: Interstitial photoradiation therapy for primary solid tumors in pet cats and dogs. Cancer Res. 41:401-404, 1981.

10. Dougherty TJ, Grindy GB, Fiel R, Weishaupt KR, and Boyle DG: Photoradiation therapy. II. Cure of animal tumors with hematoporphyrin and light. J. Natl. Cancer Inst., 55:115-121, 1975.

11. Weishaupt KR, Gomer CJ, and Dougherty TJ: Identification of singlet oxygen as cytotoxic agent in photoinactivation of murine tumor. Cancer Res. 36:2326-2329, 1976.

HEMATOPORPHYRIN DERIVATIVE IN THE DIAGNOSIS AND TREATMENT OF LUNG CANCER

Ronald G. Vincent,* Thomas J. Dougherty,*
Uma Rao,* and Daniel R. Doiron**

*Roswell Park Memorial Institute
Buffalo, NY

**Children's Hospital of Los Angeles
Los Angeles, CA

INTRODUCTION

Since 1978 we have studied the properties of hematoporphyrin derivative (HPD) and have sought means to improve the drug as well as the instrumentation used in both diagnostic and therapeutic clinical procedures (1-3).

TISSUE SPECIFICITY

In order to better ascertain the localizing properties of HPD in malignancies of the lung and esophagus, HPD was given intravenously to patients prior to surgical resection. Surgical specimens were then examined for presence of HPD and for relative intensity of fluorescence achieved when tumor tissue was exposed to violet light (405 nm). Six patients with lung cancer and one patient with esophageal cancer were injected with HPD (2.5 mg/kg) 27-96 hr prior to surgery. Upon surgical resection of the malignant lesion, the specimens were washed in saline and examined for fluorescence. Fluorescence was found in all tumor specimens upon exposure to violet light (405 nm). Additional contingent and non-contingent areas of fluorescence could be visualized when examined with the assistance of an image intensifier (Starlight Scope Tub Varo 8858).

Carcinoma of the esophagus displayed the greatest intensity of fluorescence on a scale of 1 to 3. Fluorescence was often seen as local "hot spots" especially around necrotic areas. Specimens with diffuse general fluorescence showed finger-like projections of higher intensity in cross section. Normal lung tissue showed no fluorescence when resected 72-96 hr after injection of HPD.

These data suggest that the greatest ratio of difference in HPD concentration between normal tissue and tumor tissue results when the HPD is given 48-96 hr prior to tumor resection.

FLUORESCENCE BRONCHOSCOPY

Three patients with known, previously located, squamous cell carcinoma of the bronchus were examined, as well as one patient with sputum positive for lung cancer but with negative findings on x-ray film. Dosage of HPD was 2.5 mg/kg injected 24-72 hr before in vivo bronchoscopy.

Patients were first examined by white-light bronchoscopy, then by fluorescence bronchoscopy using an instrument designed by Doiron and Profio (4). The light source was a 200 watt mercury vapor arc lamp with optics and a violet filter. An objective lens focuses an image of the fluorescent tumor on the conventional fiberoptic imaging bundle of the bronchoscope. The ocular lens at the proximal end is replaced by two lenses, a red filter to reject the reflected violet light and an image intensifier. The image intensifier is a three stage electrostatic focus, tube (Varo 8858). The image at the photocathode is converted to electrons and the electrons are accelerated and focused on a green fluorescent screen at the output of the tube. The faint red fluorescent light image is thereby transformed into a bright green image. Positive fluorescence was noted in all four patients. In patient 1, fluorescence was first visible, but was later obscured by blood. In patient 2, the tumor was highly fluorescent and showed local spots of even greater brightness. The brightness appeared to decrease slightly under prolonged illumination, possibly indicating some photodecomposition of the HPD. The tumor in patient 3 was particularly interesting because fluorescence was seen to extend farther along the bronchus than could be distinguished as tumor under white-light bronchoscopy. We believe this to be an actual extension of the tumor, although not proved by histologic examination.

Examination of the bronchial tree of the fourth patient

under conventional white-light bronchoscopy identified three suspicious areas. Two had slight color changes and the third had an altered shape, possibly abnormal. One of the discolored areas and the altered-shape area were brushed for cytologic examination, but results were negative. A biopsy could not be performed because the patient was being stressed and we decided to proceed immediately to the fluorescence examination. Positive fluorescence was observed in the two areas of discoloration, while the altered-shaped area displayed no fluorescence. Without a confirming biopsy, however, we cannot be certain the occult tumor was localized.

The fluorescence bronchoscope system appears capable of localizing small bronchogenic tumors, and probably can determine the extent of larger tumors, beyond the limits discernible under conventional white-light examination. Good differentiation between normal and malignant lung tissues is achieved between 48 and 96 hr following intravenous injection of HPD; somewhat shorter times may be acceptable. Positive fluorescence was observed in advanced tumors, but additional work with patients having very small, early lesions is required to establish the ability of the method to locate carcinoma in situ. Tests on phantoms and animals indicate the present system can detect a tumor 100 μm thick at the expected HPD concentration of 1 μg/gm (5). Much greater precision is expected with new instrumentation that is now available using a generation III - ITT image intensifier with greatly improved imaging properties, particularly in the 400 nm range.

PHOTORADIATION THERAPY

Installed in our surgical suite is an 18 watt argon-dye laser which is fitted with fiberoptic attachments of sufficient length to permit passage through a fiberoptic endoscope during examination of the patient. The dye laser emits at 625-640 nm at a maximum power of 5 watts.

In selected patients with obstructing lesions of the bronchus or esophagus who are not amenable to standard forms of therapy, we have treated the obstruction with multiple exposures of photoradiation. The surface of the tumor has been exposed to the light by placing the fiberoptic directly into the lesion, so that transillumination of the tumor is achieved. It has not been difficult to produce intraluminal necrosis of the tumor and we are proceeding to document the extent of the effects of photoradiation on normal and tumor tissues.

Typical of cases treated to date are 3 patients, two with primary lung lesions and one with an extensive tracheal

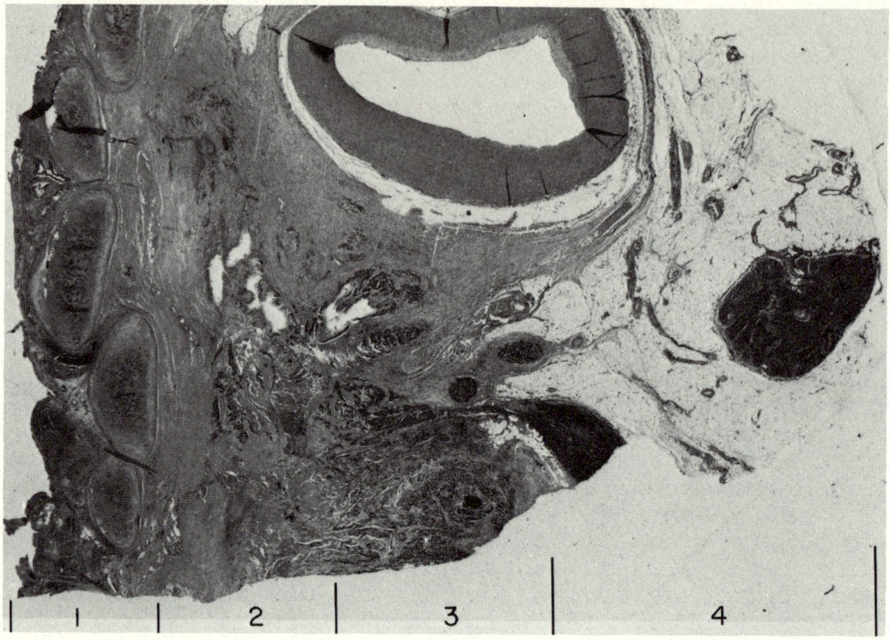

Figure 1. A whole mount tissue section of trachea that is involved with carcinoma but has been treated with photoradiation. The degree of tumor necrosis decreases as the distance from the light source increases.

Zone		Distance from Mucosal Surface
1	Regenerating Mucosa with Squamous Metaplasia, Tumor Necrosis and sloughing	0 to 2 mm
2	Tumor Necrosis with Viable Tumor	3 to 7 mm
3	Viable Tumor with Minimal Necrosis	7 to 12 mm
4	Fibro-fatty Tissue Peribronchial	12 to 20 mm

lesion. All received 2.5 mg/kg HPD, 3 or more days prior to treatment.

Patient 1 had a poorly-differentiated adenocarcinoma of the orifice of the left upper lobe. Over a period of 174 days, the patient received seven treatments which involved both a surface exposure and an optic-fiber implant directly into the tumor. The estimated exposure for each treatment was approximately 300 mW for 30 min. interstitially and 100 mW/cm² for 20 min. externally. Upon autopsy 50 days following the last photoradiation treatment, sections were taken of the tumor at varying distances from the site of insertion of the optic fiber. Viable tumor was present in all sections. Fibrosis was most extensive in areas less than 3 cm from the center of the lesion and the point of insertion of the fiber. Necrosis was present only at the mucosal surface of the tumor. Death was due to cachexia secondary to brain metastasis and a superimposed bronchial pneumonia.

Patient 2 had a large cell carcinoma of the right main stem bronchus and was treated three times over a period of 30 days with an estimated exposure of 500 mW for 20 minutes. Death occurred secondary to a large tumor that was unresponsive to treatment, metastatic disease and bronchial pneumonia. Multiple tissue sections taken at varying distances from the site of PRT demonstrated that there was tumor necrosis and fibrosis. The degree of fibrosis was more extensive at 0.5 cm than 3 cm. Viable tumor, however, was present in all sections.

Patient 3 had extensive squamous cell carcinoma of the trachea with focal extension into the esophagus and metastases to the peritracheal lymph nodes. Total occlusion of the trachea was imminent due to secretions and the size of the intraluminal component of the tumor. HPD (350 mg) was given and three days later 150 mW/cm² for 20 minutes was delivered to the surface of the tumor by a 360° diffusing monofilament fiber. Six days later, the patient died of cardiac arrest secondary to complete obstruction of the trachea secondary to an organized mucopurulent cast. Pathological sections of the treated area were made which revealed tumor necrosis up to 1 cm of depth with surface regeneration of the trachea mucosa. Viable tumor as well as degenerating cells was apparent at depths greater than 2 cm from the trachea mucosa (Fig. 1).

As more is learned about laser dosimetry and the utilization of HPD, we are hopeful that palliation if not control of these debilitating obstructive lesions will become possible.

REFERENCES

1. T.J. Dougherty, J.E. Kaufman, A. Goldfarb, K.R. Weishaupt, D. Boyle and A. Mittleman, Photoradiation Therapy for the Treatment of Malignant Tumors, Cancer Res. 38:2628 (1978).

2. T.J. Dougherty, G. Lawrence, J.E. Kaufman, D.G. Boyle, K.R. Weishaupt and A. Goldfarb, Photoradiation in the

Treatment of Recurrent Breast Carcinoma, J. Natl. Cancer Inst. 62:231 (1979).

3. T.J. Dougherty, G.B. Grindey, R. Fiel, K.R. Weishaupt and D.G. Boyle, Photoradiation Therapy II. Cure of Animal Tumors with Hematoporphyrin and Light, J. Natl. Cancer Inst. 55:115 (1975).

4. A.E. Profio, D.R. Doiron and E.G. King, Laser Fluorescence Bronchoscope for Localization of Occult Lung Tumors, Med. Phys. 6:523 (1979).

5. D.R. Doiron, E. Profio, R.G. Vincent and T.J. Dougherty, Fluorescence Bronchoscopy for Detection of Lung Cancer, Chest 76:27 (1979).

LASER PHOTORADIATION THERAPY OF RECURRENT HUMAN BREAST CANCER AND CANCER OF THE HEAD AND NECK

A. G. Wile, A. Dahlman, R. G. Burns,
G. R. Mason, F. M. Johnson* and M.W. Berns**

Department of Surgery
University of California
Irvine, CA

*Department of Physics
California State University
Fullerton, CA

**Department of Developmental and Cell Biology
University of California
Irvine, CA

INTRODUCTION

We have been involved in a clinical trial of photoradiation therapy (PRT) for recurrent cancer. This report describes our initial experience with this technique. The aim of this pilot study was to determine practicality of PRT. Once this aim had been accomplished, our next goal was to refine treatment parameters to maximize the theoretical selective cytotoxicity of hematoporphyrin derivative (HPD) and light (625-640 nm).

MATERIALS AND METHODS

Patients. Patients had diagnosed cancer recurrent to the skin, subcutaneous tissue, or mucous membranes of the oropharyngeal cavity. These tumors were refractory to all modes of conventional therapy. Patients were informed that PRT is an experimental therapy, not an alternative to con-

Table 1. Tumor response to PRT with respect to histology.

Histology	No. of Patients	No. of Lesions	CR	PR	SD	NR
HEAD AND NECK*	6	7	3	1	0	3
COLON	1	2	0	0	2	0
CERVIX	1	1	0	0	0	1
BREAST*	10	34	12	18	2	2

* Patients with 14 lesions were excluded because of other concomitant medical therapy. Two head and neck and 3 breast cancer patients received 2 cycles of therapy.

ventional cancer management. They were instructed to avoid sun exposure after injection of HPD to avoid a skin reaction due to the heightened photosensitivity.

This report describes our experience with 18 patients. The distribution of the sites of primary are as follows: Head and Neck - 6, Colon - 1, Cervix - 1, Breast - 10. Five patients received a second injection of HPD and were treated with PRT for a second cycle. HPD was supplied by Thomas J. Dougherty (Roswell Park Memorial Institute). It was injected intravenously as a single bolus over several minutes. The doses ranged from 3 to 5 mg/kg of body weight.

Light delivery system. Red light at 630 nm was utilized to activate the photochemical reaction resulting in cell destruction. The system utilized was an Argon-pumped dye laser coupled to a plastic clad optical quartz monofilament fiber. Initial treatment was accomplished with a 4 watt Argon laser (Spectra-Physics Model #164). Subsequently this laser was replaced with a 12 watt Argon laser (Spectra-Physics Model #171) that was used to pump a dye laser (Spectra-Physics Model #375). The circulating dye was Rhodamine B (0.79 g/l, Exciton Inc.). The dye solution was changed after 16 hr of continuous use. The fiber optic was coupled to the dye laser by means of a micrometer adjusted alignment device (Spectra-Physics Corp.).

Light was applied to the tumor bearing area by superficial illumination or by implantation of the fiber into the tumor mass. Implantation was utilized in 3 patients.

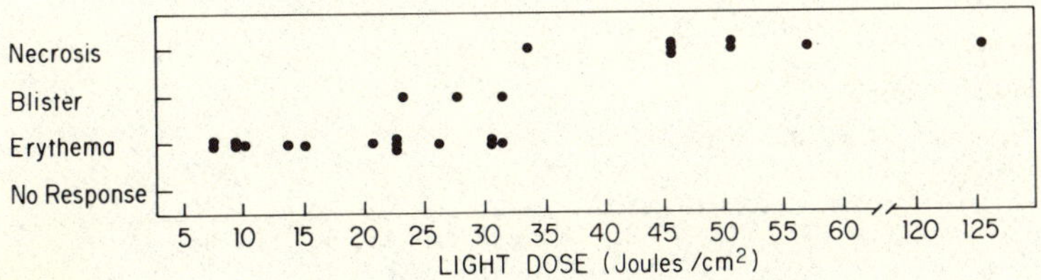

Figure 1. Relation between total light dose vs. response. Doses of light between 8 and 20 joules/cm² were associated with good clinical response and only erythema of adjacent tissue.

Treatment protocol. Patients were injected with HPD on day 1. They were then allowed to leave the outpatient area, were cautioned to remain indoors, and returned 72 hr later for photoradiation therapy. The area to be treated was encircled with indelible ink. The intensity of light emanating from the end of the fiber was measured using a power meter (Spectra-Physics Model #404). By knowing light intensity and the surface area to be treated, the dose rate and total dose of light were calculated and recorded for each treatment area.

Three patients underwent intratumor placement of the fiber optic. The 400 micron diameter fiber was passed directly into the tumor or was fed through a 15 gauge needle introduced with aseptic technique under the skin.

Patients returned 24 hr after therapy for evaluation of response and for irradiation of areas that were not treated on the previous day due to time constraints. Patients then returned 1, 2 and 4 weeks post therapy for evaluation of response. Five patients (2 Head and Neck and 3 Breast cancer) underwent a second cycle of therapy.

Ultimate tumor response was judged only at 4 weeks post therapy. Response was graded by standard criteria: complete response /CR/ = resolution of all tumor, partial response /PR/ = reduction by 50% or more of the product of the cross sectional diameters, stable disease /SD/ = reduction in tumor mass but not to the point of a partial response, and no response /NR/. Photographs were taken throughout the course of therapy to further document response.

RESULTS

The ability to eradicate recurrent tumor was generally very good. Three patients with 14 lesions treated were ex-

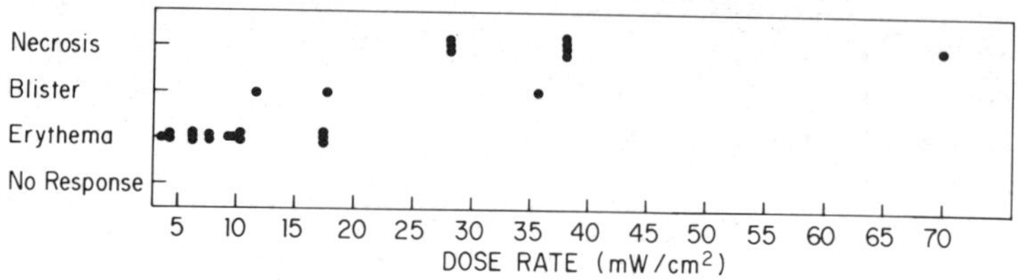

Figure 2. Relation of rate of administered light to response.

cluded from the overall evaluation because they received concomitant chemotherapy; all three had progressive disease despite chemotherapy and yet had response rates (CR + PR) in excess of 90 %. The remaining 15 patients had a total of 44 lesions treated. The overall response rate was 77% (34 CR + PR, Table 1).

Toxicity of HPD was minimal. There was no acute reaction to the injection of the HPD. Photosensitization of normal skin was a problem. Most patients were able to avoid skin reaction with heavy clothing and hats during trips to the medical center. Several patients did receive enough sunlight exposure to cause erythema and edema of exposed skin. This reaction was transient and always resolved within 24 to 48 hr. A single instance of skin infiltration of HPD during drug injection occurred. Heightened sensitivity to sunlight was noted in this area but there were no other serious consequences.

Responses were apparent at 24 hr. Selective tumor necrosis was readily observed with lesser changes in the surrounding tissue. Necrosis of large tumor masses resulted in ulceration whereby healing occurred by scab formation with underlying epithelialization or by granulation tissue.

Patients with breast cancer provided a group large enough to evaluate separately. Thirty of 34 lesions responded favorably (88% CR +PR). Light dose (joules/cm^2) and dose rate (mw/cm^2) were varied by altering treatment time and surface area. High total doses of light (>20 joules/cm^2) and high dose rates (>10 mw/cm^2) were associated with blistering and necrosis of normal skin and tumor tissue (Figs 1,2). Total light doses of 8-20 joules/cm^2 evoked erythema in normal tissue but gave excellent therapeutic results; doses less than 8 joules/cm^2 caused no effect on normal or malignant tissue. When comparing total light dose and dose rate, it became apparent that in our treatment modality these variables were interrelated (Fig. 3).

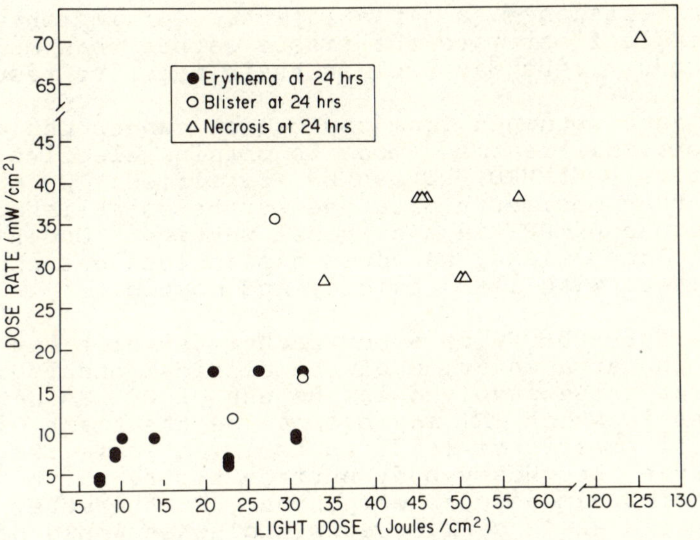

Figure 3. A comparison of total light dose vs. dose rate in our clinical protocol. In this study, high dose rates were always associated with large total light doses.

The three patients undergoing intra-tumor placement of the fiber optic had no response to therapy. The last patient in this group was treated the most vigorously. A tumor 3 cm in diameter was implanted with a fiber emitting 750 mw of light for a period of 30 minutes.

DISCUSSION

This initial effort involving PRT at our institution resulted in response rates comparable to those reported by other investigators in this field (1). We found that we were able to achieve greatest success with those tumors that were flat rather than elevated bulky tumors. Although our experience with colon and cervical cancer is too small from which to draw conclusions, the results with head and neck cancer and breast cancer suggest that neither histologic appearance, nor success or failure of prior therapy are indicators of response to PRT. All tumors appear to respond equally well.

We have been less successful with the technique of fiber implantation. Based upon our results with superficial illumination the patients described above should have had a significant response to implantation. Our implantation of only a single fiber into the tissue rather than multiple simultaneous fibers may explain these negative results.

The data obtained from the breast cancer patients suggest an optimal treatment mode to obtain selective tumoricidal activity of HPD-PRT can be identified. We believe strongly that dosimetry data need to be established for appropriate use of PRT in a clinical setting. Under optimal dosimetric conditions, we would expect less destruction of normal tissue with less toxicity and morbidity.

Clearly, HPD-PRT is a new technique which has gained a place in the armamentarium of the clinical oncologist. It remains for those involved in the use of PRT to determine situations in which PRT may become the treatment of choice. Our initial experience leads us to investigate the treatment of primary or secondary body surface tumors which have a minimal third dimensional component. Furthermore, small macroscopic disease or microscopic disease would be expected to respond most favorably to this therapy.

ACKNOWLEDGEMENTS

Part of this research has been conducted with facilities of the NIH LAMP Biotechnology Resource supported by Grant RR 01192, and from NIH Grant GM23445 to Michael W. Berns.

REFERENCE

1. Dougherty, T. J. Photoradiation therapy for cutaneous and subcutaneous malignancies. J. Invest. Dermatol. 77:122-124, 1981.

HPD PHOTORADIATION THERAPY FOR CANCER AT KINGSTON AND HAMILTON

James Kennedy

Ontario Cancer Foundation
Kingston Clinic
Kingston, Ontario Canada

The Table accompanying this article summarizes our experience with photoradiation therapy from April 1978 to the present. Evaluation of relative responsiveness was based on the following considerations. A 5+ rating means that the tumor can be destroyed by a dose of light that will cause only mild erythema and slight edema; a 3+ rating indicates that a moderate amount of skin damage may be expected from the minimum dose of light required to destroy a subcutaneous nodule; a 2+ rating means that the degree of skin damage to be expected will be on the border between acceptable and not acceptable; and a 1+ rating indicates that serious skin damage is inevitable if a subcutaneous nodule is to be treated successfully by external beam. These evaluations of relative responsiveness are only approximate, since the response of a tumor mass to treatment depends upon the volume of the mass and its depth within or below the skin as well as upon its histology. Since all of the basal cell carcinomas we have treated have been relatively superficial, and since small and superficial intracutaneous nodules of breast carcinoma respond very well to relatively low doses of photoactivating light, the responsiveness of breast carcinoma in absolute terms may well be very similar to that of basal cell carcinomas.

Our standard dose of HPD was 2.5 mg/kg of body weight, our standard interval between injection and photoactivation was either 4 or 5 days, and our standard source of photoactivating light was an old style Kodak Carousel slide projec-

TYPE OF MALIGNANCY	PATIENTS TREATED	RELATIVE RESPONSIVENESS
basal cell carcinoma	3 (total of 38 primaries)	5+
squamous cell carcinoma		
(a) of skin	4 (multiple secondaries)	2+
(b) of mucosa	4 (multiple secondaries)	2+
malignant melanoma		
(a) heavily pigmented	3 (multiple secondaries)	1+
(b) lightly pigmented	1 (multiple secondaries)	3+
carcinoma of breast	16 (multiple secondaries)	3+
carcinoma of endometrium	1 (multiple secondaries)	3+
carcinoma of ovary	1 (single nodule)	3+
carcinoma of lung (squamous cell)	1 (single nodule)	could not evaluate
Kaposi's sarcoma	2 (multiple secondaries)	3+
retinoblastoma	1 (multiple secondaries)	3+

tor equipped with a 500 watt tungsten lamp and the usual condensing lenses and heat filter, and carrying a 50 mm square Hoya R-60 sharp-cut colour filter in the slide compartment. The output from this light source has a biological activity approximately half that of the output from an untuned dye laser loaded with Rhodamine B. A dose of 25 mWhr/cm^2 (90 joules/cm^2) will cause no more than mild erythema and slight edema of the skin of most patients (and will destroy superficial basal cell carcinomas), a dose of 30 to 35 mWhr/cm^2 (108 to 126 joules/cm^2) will cause strong erythema and substantial edema of the skin, with dry desquamation and perhaps slight blistering in especially sensitive skins, while doses of 40 mWhr/cm^2 (144 joules/cm^2) or more are often associated with the formation of deep blisters and eventual scarring of the skin. Consequently, if the treatment field is large and includes normal skin, we rarely give above 35 mWhr/cm^2. However, treatment fields smaller than approximately 1 cm in diameter can tolerate doses up to 60 mWhr/cm^2 (216 joules/cm^2) without serious damage. Vaginal mucosa also can tolerate doses of up to 60 mWhr/cm^2 if the fields are small. If the malignant tissue has penetrated through the skin to such an extent that the photoactivating light need not be transmitted through skin, we will give approximately 200 mWhr/cm^2 (720 joules/cm^2) for every cm of tumor thickness. When we use a plastic rod to transmit the light directly into the center of an accessible mass of malignant tissue, we generally give a total dose of 700 joules (194 mWhr) to produce a zone of killing 1 to 2 cm in diameter.

Our clinical experience with HPD photoradiation therapy includes the following histological types of cancer.

Basal cell carcinoma. We have treated 38, ranging in size from a very superficial tumor 2 mm in diameter to a lesion with a central ulceration approximately 3 mm deep and a total diameter of approximately 12 mm. All vanished in response to doses of light that caused only mild erythema and slight edema. The cosmetic results were excellent, far better than what is usually seen following low voltage X-ray therapy. There was no depigmentation and no loss of either hair or subcutaneous tissue in the treatment fields. There have been no recurrences up to the present; our longest follow-up, for the largest lesion described above, has been 35 months. Premalignant lesions also have responses to photoradiation therapy.

Squamous cell carcinoma. We have treated 8 in total, 4 originating in the skin and 4 in the mucous membranes of the mouth or nose. The results have been rather disappointing. These tumors certainly can be killed by means of photoradiation therapy, but their sensitivity is not very much greater than that of normal skin, and their habit of producing a mass of closely-packed subcutaneous nodules with necrotic centers makes it very difficult to kill the malignant tissue without also destroying the blood supply to the overlying skin. We managed to achieve histologically-verified local control in one patient, but he died shortly afterwards of a brain secondary. It is perhaps worthy of note that this man's tumor involved mucous membrane rather than skin, since mucous membranes in general appear to be less easily damaged than is skin by a given dose of photoactivating light.

Malignant melanoma. Our experience is similar to that reported by others. Even quite large nodules can be destroyed if the tissue is only lightly pigmented, but large heavily-pigmented nodules are practically impossible to treat effectively, even when optical fibers are inserted deep to their base.

Breast carcinoma. We have treated 16 patients with multiple cutaneous and subcutaneous metastases, and have come to the following conclusions. The ideal situation is a patient with one or more small cutaneous nodules involving a relatively small area of the chest wall. Next best are small subcutaneous nodules restricted to a small area. A thick mass of malignant tissue restricted to an area perhaps 10 x 10 cm is still a reasonable proposition. Multiple small cutaneous or subcutaneous nodules involving a large area may respond to photoradiation therapy, but any benefit obtained appears to be of very short duration. The results of treating patients with massive involvement of large areas of the chest and back are disappointing. It is possible to kill the tumor down to bare ribs, but the patient does not really benefit from this procedure. In summary, we try to treat when the lesions are small, few, superficial, and restricted to a relatively small area, and avoid treating massive lesions involving a large area. Consequently, as originally suggested by Dr. Dougherty, it makes good sense to give a second treatment about 3 weeks after the first has reduced the tumor masses to a more manageable size.

Local control is achieved most easily if the tumors have a long doubling time, either because of the nature of the tumor:host interaction or because of concomitant systemic therapy. In principle, it makes sense to use photoradiation therapy in combination with either hormonal manipulation or chemotherapy. However, be extremely careful if the patient has sometime in the past been given both ionizing radiation and Adriamycin. We treated one such patient following radiotherapy but prior to the administration of Adriamycin, and found that the skin of the irradiated and non-irradiated areas of her chest wall were indistinguishable in their reactions to our standard dose (30 mWhr/cm^2) of photoradiation. She was later given

Adriamycin, and subsequently returned for additional photoradiation therapy. This time we found that a dose of 5 mWhr/cm² given to a previously irradiated area produced symptoms (itching and tingling) that are usually found only after much larger doses, and also produced the same intensity of erythema as did a dose of 30 mWhr/cm² that was given to an area that had never been treated by ionizing radiation. Cautious testing of other patients who had received both ionizing radiation and Adriamycin at various times prior to photoradiation therapy produced results indicating that an effect of Adriamycin on previously irradiated skin persisted, with decreasing severity, for at least several months following cessation of the chemotherapy. We have not yet found any evidence of a similar type of interaction between photoradiation therapy and other chemotherapeutic agents.

The skin of patients with inflammatory breast carcinoma can react very strongly to photoradiation therapy. Itching begins after a very small dose, and intense erythema develops during treatment. Both the early itching and early erythema can be controlled by giving an antihistamine 30 minutes prior to treatment. A normal degree of erythema is seen a few hours later when the effect of the antihistamine wears off. I would advise the use of antihistamines during photoradiation of patients with inflammatory breast carcinoma, not just for their comfort but also to prevent the severe skin reactions that may be anticipated in any area of skin inflammation.

We encountered one extremely fascinating problem during our treatment of chest wall breast disease. The nodules of malignant tissue in our 16 patients all showed about the same response to a given dose of light, with one exception. We had a patient with multiple nodules, all but one of which showed the usual sensitivity. That one nodule was completely resistant to photoradiation therapy. It first appeared in an area that we had treated previously by photoradiation therapy, and then grew much more rapidly than did any of the other nodules. Histologically, it was indistinguishable from the other nodules. It had a good supply of well-oxygenated blood and did not show any area of central necrosis. The first time we tried to kill it by means of photoradiation therapy, we gave it a total of 585 mWhr/cm² (2106 joules/cm²) on days 4 to 6 post-injection. This dose of light, approximately 20X larger than our standard dose for chest wall nodules, killed this particular nodule to a depth varying between 0.5 cm and 1.0 cm at different sites. On our second attempt one month later, we gave a total of 926 mWhr/cm² (3330 joules/cm²) over post-injection days 4 to 7, and in addition we injected an extra 25 mg of HPD directly into the tumor 30 minutes before each treatment. This drastic treatment resulted in the killing of tissue only within the outermost 2 mm of the nodule, and it appeared to me that even this was not the result of photoactivation but rather an effect of hyperthermia and dehydration of the outermost layers. Other nodules treated at this point in time were destroyed by as little as 15 mWhr/cm^2 (54 joules/cm^2). It appears then that our resistant nodule was the result of a mutation, its increasing resistance with repeated treatment being the natural result of selective destruction of cells of normal sensitivity which were present in the original mixed culture.

The question arises as to whether or not the mutation arose as a result of previous photoradiation therapy, since the nodule appeared in a previously treated area. In theory, a toxic agent that kills cells by oxidizing their membranes might, very rarely, damage the DNA of a cell without also causing enough membrane damage to kill that cell. When we looked into this possibility by means of the Ames test (bacterial mutagenesis), using three different types of bacteria with and without liver extract and with and without the presence of photoactivating light, we could not find any clear evidence that HPD was mutagenic under any circumstances. However, it would be well to keep in mind the possibility that photoradiation therapy might cause mutations and therefore may be slightly carcinogenic.

Endometrial carcinoma. We have treated one patient with an inoperable vaginal vault recurrence of endometrial carcinoma, by going in through the vagina with either a cone or a light pipe. At first we were very cautious and gave rather small doses because of the proximity of the bladder and the rectum, so the tumor always recurred. Lately we have been increasing the dose, our last being 700 joules given into three different sites. We have now treated this patient 20 times over a period of 28 months. Her general health is excellent, and we have seen no evidence whatever of any cumulative toxicity from the repeated injections of HPD. Experiments carried out in animals by Dr. A.A.F. Sima et al. (1) indicate that HPD can cause a transient neurotoxicity. These *in vivo* observations are supported by the results of *in vitro* studies carried out by R. Riopelle et al (2). We must therefore keep in mind the possibility that repeated injections of HPD may have a cumulative effect on the nervous system, and keep alert for possible clinical manifestations of neurotoxicity.

Carcinoma of the ovary. Our one patient had an inoperable vaginal vault recurrence. We treated her exactly as we did the patient with the endometrial carcinoma, starting with low doses for fear of damage to the bladder or bowel and then working up. Our latest dose was 700 joules, directed into the tumor. Our only problem so far has been a tendency of the patient to bleed slightly following treatment and a fear of the physician that we will find ourselves in communication with the peritoneal cavity if we manage to kill all of the tumor.

Kaposi's sarcoma. We used three different treatment techniques for these patients. To prevent the development of new lesions, we gave the whole of the lower leg a dose of 15 mWhr/cm² (54 joules/cm²). To treat small cutaneous lesions (up to 5 mm in diameter), we used small fields limited by a cone and gave between 25 and 40 mWhr/cm² (90 to 144 joules/cm²). The massive lesions of advanced disease were palliated by the use of much larger doses given directly into the malignant tissue.

Retinoblastoma. Superficial retinoblastoma secondaries within the orbital cavity responded well to doses of 20 to 25 mWhr/cm² (72 to 90 joules/cm²). The secondaries vanished and the tissue became epithelialized. A single subcutaneous secondary was partially destroyed by a dose of 35 mWhr/cm² (126 joules/cm²).

LIGHT SOURCES AND LIGHT DELIVERY SYSTEMS

-Rhodamine B dye laser, untuned. We use our laser only if the clinical situation is such that light delivery via optical fiber is the best or only option. For example, much of the nasal cavity is most easily illuminated via an optical fiber, and the use of an optical fiber is essential if you are trying to illuminate the base of a large, heavily-pigmented melanoma nodule. However, most of our patients are treated by means of low technology light sources. The dye lasers presently available do not have the output required for the treatment of a large area, and we find the simpler technology much more convenient as well as much less expensive when repairs become necessary.

Our standard light source for treating areas up to approximately 10 cm in diameter is an older model Kodak Carousel slide projector equipped with a 500 watt tungsten lamp and the standard condensing lenses and heat filter. We put a Hoya R-66 glass filter in the slide compartment when lining up on our target (the light passed by this filter does not photoactivate HPD), and then replace it by a Hoya R-60 filter for treatment. Although the newer model Carousel projectors (Series "H") with quartz-halogen lamps can be used, they do not produce as much power in the spectral region of interest as do the older models with the tungsten lamps. The output of the tungsten lamps can be increased (at the expense of lamp life) by using a higher voltage. It is possible to replace the standard DEK lamp with a DEL lamp to obtain approximately 20% greater output, at the cost of a shorter lamp life.

For treatment fields up to approximately 3 cm in diameter, we replace the projection lens by a cone made of semi-reflective aluminum foil. The total output varies from approximately 600 mW when the exit diameter of the cone is 0.6 cm, to approximately 1750 mW when the exit diameter is 3.0 cm. If the tip of the cone is placed directly against the tissue to be treated, the distribution of light is surprisingly uniform over the whole area. However, focussing effects occur if the tip is separated from the target tissue. Quite recently we have been evaluating a conical plastic light guide as a possible replacement for the foil cones. It can produce an intensity of over 5000 mW/cm^2 in a field 0.5 cm in diameter, enough to cause pain or to char clotted blood if contact is not maintained with the heat sink provided by a large mass of tissue.

For fields between 3 and 10 cm in diameter, we generally use the projector along with a standard 5" f/2.8 projection lens. The use of lenses of higher f-numbers greatly reduces the output, and the use of lenses of shorter focal length increases the curvature of the isointensity curves. It is important to realize that <u>all</u> projection lenses produce a higher intensity at the center of a flat field than they do at the edges. Consequently, we routinely use only the central 2/3 of a beam.

Cones can be used to introduce light directly into a tumor, if the tip of the cone is of small enough diameter. The conical plastic light guides are of special value in this regard, since the end of the guide can be extended as a rod of appropriate diameter. It is a good idea to turn off the projector before removing such a rod from a tumor, since blood or tissue adhering to the tip of the rod quickly becomes charred and may absorb enough energy to melt the tip of the rod. Since output is increased by approximately 20% if the tip of the rod is in contact with water, it makes good sense to coat the tip with material of an intermediate refractive index and to make suitable corrections to the intensity measurements if these are made in air.

For fields larger than 10 cm in diameter, we use a light source consisting of four 1000-watt high-pressure sodium lamps. These are placed in relationship to each other in such a way that the intensity of illumination of a flat surface placed 14 cm away from the near surface of the filter averages 100 mW/cm^2 and shows less than ±2.5% variation over a rectangular area measuring 24 cm x 23 cm. The light is filtered to produce approximately the spectrum transmitted by a Kodak Wratten 26 gelatin filter or a Hoya R-60 glass filter. We pass the light through an aqueous solution containing Rhodamine B (8 x 10^{-5} g/ml) and Tartrazine (2 x 10^{-3} g/ml) which is continually recirculated between the Pyrex glass windows of the filter and a water-cooled heat exchanger. Recently, because of problems in obtaining Tartrazine that is chemically and optically equivalent to our original batch, we have been using an aqueous solution of Rhodamine B (0.2 mg/ml) as the circulating filter material, plus a sheet of red acetate film (Transilwrap of Canada) fastened to the outer surface of the glass window. We generally use the high pressure sodium source for treating large fields involving the chest wall or the back, or occasionally for treating a limb. At a dose rate of 100 mW/cm^2, a typical treatment is 15 to 20 minutes.

It should be noted that a power meter designed for use with lasers can not be used for accurate measurement of the output of this high pressure sodium source, since the illumination produced is very diffuse. We use a cosine-corrected black-body Eppley Radiometer, but any cosine-corrected black body meter of adequate sensitivity could be used.

ARE WE READY FOR RANDOMIZED TRIALS?

Patients would be randomly assigned to either HPD photoradiation therapy or the best available alternative therapy, including no therapy. Do we know enough about the probable benefits and hazards of photoradiation therapy in certain specific clinical situations to be able to honestly state that photoradiation therapy is at least as likely as any conventional therapy to produce significant benefit for the patient? Consider the following possibilities.

(A) As the primary treatment for multiple primary basal cell or squamous cell carcinomas that are too numerous and too widespread to be reasonably treated by either radiotherapy, standard surgery, electrodesiccation, or cryosurgery, and too resistant for systemic chemotherapy.

Alternatives

photoradiation therapy
vs
topical chemotherapy (squamous cell carcinoma)
or
no treatment (basal cell carcinoma)
or
topical immunotherapy (either)

(B) As localized adjuvant therapy given immediately following attempted curative surgery for malignant melanoma.

Alternatives

photoradiation therapy
vs
no treatment
or
experimental adjuvant chemotherapy
or
experimental adjuvant radiotherapy
or
experimental adjuvant immunotherapy

(C) As a palliative treatment for breast carcinoma recurrent to the chest wall in patients who have been treated in the following sequence: surgery ± adjuvant chemotherapy or adjuvant radiotherapy for the primary, followed by tissue tolerance doses of radiotherapy for the first chest wall recurrences.

Alternatives

photoradiation therapy
vs
hormonal manipulation
or
systemic chemotherapy

In all three examples, it is intended that there be only two treatment arms to the clinical trial, the specific alternative to photoradiation therapy being chosen according to local preference. Failure of one treatment arm in a given patient would be an indication to cross over to the other arm.

Are we really ready for phase III clinical trials such as the three briefly described above? I suggest that we are. We now have enough experience with HPD photoradiation therapy to be able to evaluate the probable risks and probable benefits for any patient who might be eligible for any of the above trials. Shall we give it a try?

REFERENCES

1. A.A.F. Sima, J.C. Kennedy, D. Blakeslee, and D.M. Robertson, Experimental Porphyric Neuropathy: A Preliminary Report, Can. J. Neurol. Sci. 8:105 (1981).

2. R.J. Riopelle and J.C. Kennedy, Analysis of Porphyrin Toxicity *in vitro*, Can. J. Physiol. Pharmacol. In Press.

LIGHT DOSIMETRY IN TISSUE: APPLICATION TO PHOTORADIATION THERAPY

Daniel R. Doiron*+, Lars O. Svaasand†, and A. Edward Profio*‡

*Institute of Physics and Imaging Science
University of Southern California
School of Medicine
Los Angeles, California 90027

+Clayton Ocular Oncology Program[1]
Children's Hospital of Los Angeles
Los Angeles, California 90024

†Division of Physical Electronics
University of Trondheim
Trondheim, Norway

‡Department of Chemical and Nuclear Engineering
University of California
Santa Barbara, California

INTRODUCTION

Dosimetry in photoradiation therapy (PRT) is an area in need of research and development. If PRT is to be optimized and perfected for routine clinical use, methodology and techniques for predicting and monitoring the therapeutic reaction, e.g., dose, need to be developed. The ultimate effectiveness of the photodynamic reaction is a function of the concentration of the photosensitizing agent HPD, and the effective light flux, at a given spatial point within the

---

[1]Address communications to Dr. Doiron at this address.

Table 1. Factors Affecting the Therapeutic Dose in PRT

PRT effectiveness α(Light Flux) (Drug Concentration)

I. <u>Light Flux</u>
   1. Spectrum
      a) Coupling to drug absorption spectrum <u>in vivo</u>
   2. Attenuation coefficient of tissue
      a) Wavelength dependence
      b) Tissue properties
         (1) pigmentation
         (2) heme content
         (3) absorption and scattering

II. <u>Drug Concentration in vivo</u>
   1. Delivered drug dose (mg/kg)
   2. Effectiveness of the drug
      a) Localization
      b) Photosensitization
   3. Delay Time after injection
   4. Cellular and tissue properties

tissue. Factors affecting these two parameters are outlined in Table 1. In order to know accurately the dose, the drug concentration, light radiance and effective interaction of these quantities must be determined. Physiological conditions of the tissue may also play an important role in the therapeutic response obtained. The oxygen content may be an important factor if singlet oxygen is the major reactive molecule in PRT (1).

It has been the object of this work to theoretically and experimentally evaluate light penetration and distribution in tissue to aid in the determination of this contribution to the therapeutic dose. Little is known of the <u>in vivo</u> concentration of HPD localizing component(s); this subject is not included in this work. The effective light dose at a given spatial point in tissue is determined by the light spectrum and the absorption of the photosensitizing component. The dose at depth in a tissue, referred to as depth-dose, is determined by the delivered irradiance and the optical absorption and scattering characteristics of the tissue and the photosensitizing agent for the wavelength of light used.

LIGHT DISTRIBUTION IN TISSUE

The optical characteristics of tissue have been published at a limited number of wavelengths and for few

Table 2. Light distribution in tissue: absorption-dominant and diffusion (scattering)-dominant conditions.

	ABSORPTION DOMINANT	DIFFUSION DOMINANT
1-Dimension	$I = I_o e^{-\alpha x}$	$\phi = \phi_o e^{-\alpha x}$
3-Dimension (Spherical Symmetrical)	$I = I_o \left(\frac{a}{r}\right)^2 e^{-\alpha(r-a)}$	$\phi = \phi_o \left(\frac{a}{r}\right) e^{-\alpha(r-a)}$
α : Attenuation Coefficient:	$\alpha = \beta + \kappa$	$\alpha = \sqrt{\frac{\beta}{\zeta}}$

β: Absorption Coefficient

κ: Scattering Coefficient

ζ: Diffusion Coefficient $\cong 1/3\ \kappa$

I: Power Density

ϕ: Space Irradiance

a: Source Radius

specific tissue types (2,3). Except for ocular structures and skin, the data are mainly given for wavelengths related to the surgical use of lasers: 488 and 514.5 nm for argon lasers, 1.060 microns for the Nd-Yag laser and 10.6 microns for the CO_2 laser. A more detailed study is needed at the wavelengths utilized in PRT.

Most tissue structures are highly inhomogeneous, making the exact solution of the equations governing the propagation of light (Maxwell's Equations) impractical. This inhomogeneous structure permits the neglect of spatial coherence and interference mechanisms of the light and permits substantial approximations to be made. Under such conditions, the laws of geometrical optics can be used. Table 2 outlines solutions using such approximations for the light distribution in a medium where either absorption or scattering mechanisms are dominant. These solutions were derived from conservation of energy principles using a diffusion theory approximation in the case of the scattering-dominant condition (4). The solutions differ only by the additional 'a/r' factor for the spherically symmetrical absorption-dominant case and the definition of the attenuation coefficient. In diffusion theory, the attenuation coefficient (α) is related to the square root of the ratio of the absorption coefficient to the diffusion coefficient; for the absorption dominant case, it is the sum of the absorption coefficient + the scattering coefficient. In either case, the 'a/r' or '$(a/r)^2$' factors will only be important for values of $r=1/\alpha$. At distances greater than the penetration length ($\delta=1/\alpha$), the space irradiance is deter-

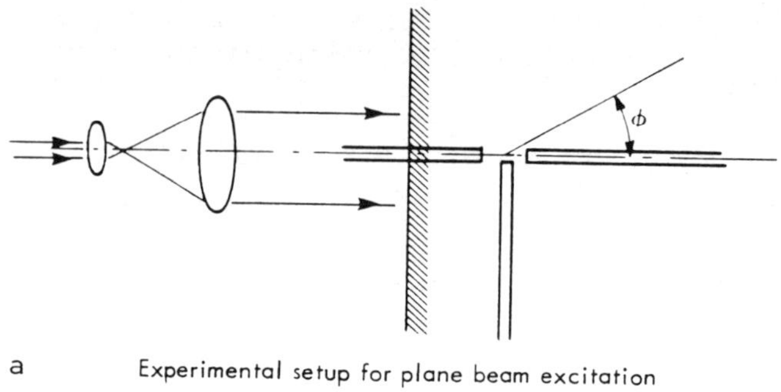

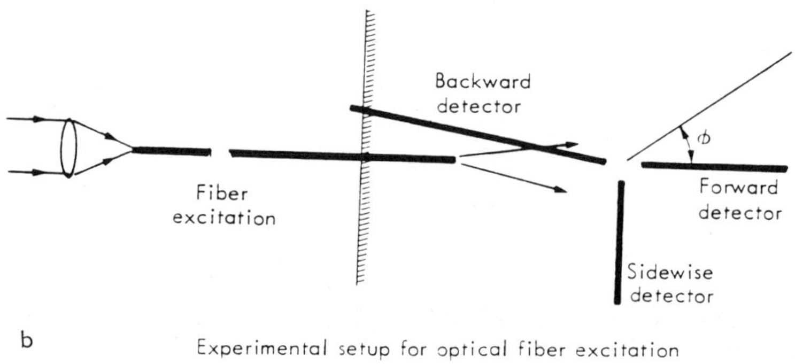

Figure 1. Representation of single fiber optical probe measurement of light distribution in tissue.

(a) - surface excitation (plane beam)

(b) - optical fiber excitation

mined by the exponential factor. The attenuation coefficient can therefore be determined by measurement of the space irradiance at points distant from the source location.

Spatial distribution of the irradiance in tissue can be mapped using single fiber optic probes and appropriate light detection systems. Fig. 1 schematically shows such a method for measuring the light in tissue illuminated by either plane beam exposure or optical fiber implantation. The output of each fiber is connected to a photomultiplier or other suitable light detector with appropriate read-out. An optical filter or monochromator between the fiber output and detector is desirable to assure rejection of light wavelengths other than those being investigated. By measuring the irradiance in the forward, backward and sideways direction, the spatial radiance can be calculated. This is done

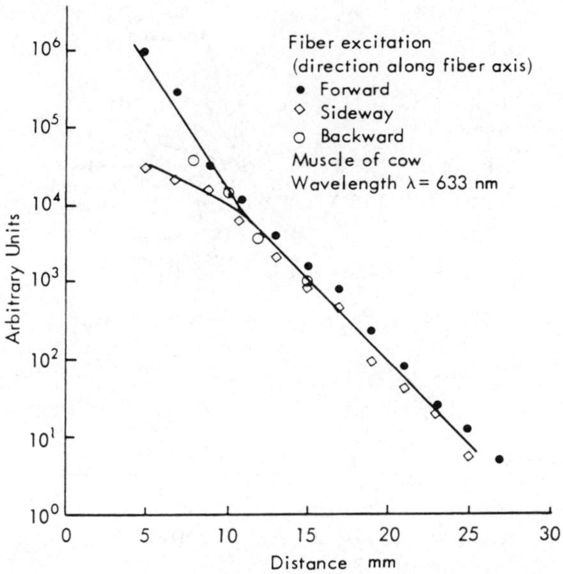

Figure 2. Radiance vs. distance from excitation surface. Measured light distribution in the forward, sideways and backward direction in cow muscle (*in vitro*) for fiber interstitial excitation of 633 nm light. Distance is from exciting fiber face.

by weighting the individual measured radiance by their appropriate solid angle acceptance and summing. Fig. 2 shows the measurement of the relative radiance values in the different directions for cow muscle *in vitro*. This data was obtained by interstitial implantation of an illumination fiber connected to a helium-neon laser, (λ =633 nm). Near the exciting fiber, the distribution is directionally dependent due to the nature of the illumination system. Beyond 10 mm, the light has become isotropically distributed and can be representatively measured by a single directional probe. In this region, the attenuation coefficient can be determined by the slope of the curve.

Weighting these different directional measurements by their relative solid angle acceptance, as shown in Fig. 3, the actual space irradiance can be obtained. Measurements for different animal tissues obtained from a commercial meat supplier are shown in Fig. 4. These data are based on surface illumination of the tissue and are normalized to the same value at 0.0 mm. The attenuation coefficient, penetration depth and reflection coefficient from these measurements are given in Table 3. The penetration depth is defined as the inverse total attenuation coefficient and

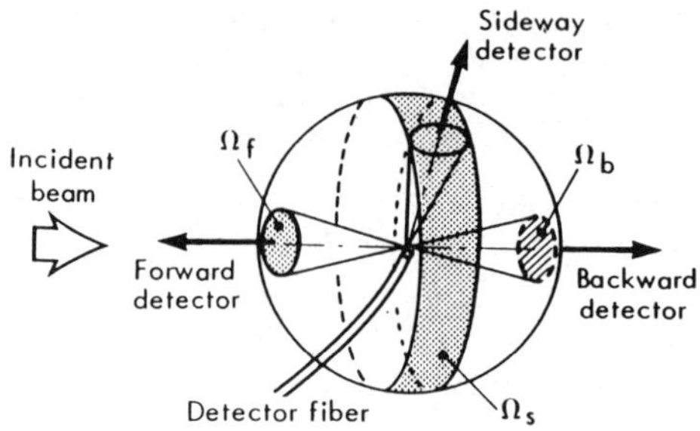

Solid angles covered by the detector fibers

SPACE IRRADIANCE

$$\phi = 4\pi \frac{[I(\Omega)_f + I(\Omega)_b]\frac{NA}{n} + 4I(\Omega)_s}{2\frac{NA}{n} + 4}$$

$I(\Omega)_f$ = FORWARD IRRADIANCE

$I(\Omega)_b$ = BACKWARD IRRADIANCE

$I(\Omega)_s$ = SIDEWAYS IRRADIANCE

NA = NUMERICAL APERTURE OF FIBER

n = INDEX OF REFRACTION OF TISSUE

Figure 3. Schematic representation of solid angles of different directional pickup fibers and equation for determining space irradiance based on individual directional irradiance measurements.

represents the distance at which the total optical power is reduced to 37%. The optical power is reduced to 10% at a distance $\ln 10\,\delta = 2.3\delta$.

Fig. 5 shows radiance vs. depth measurements in resected human lung tissue specimens at a wavelength of 633 nm. The lung substance measurements were obtained using a 200 micron core diameter silica optical fiber mounted in a 20 gauge spinal needle. The bronchial membrane and squamous cell carcinoma specimen required a variation of the fiber implantation technique due to their limited thickness. Fig. 6 shows the method used to measure light attenuation. The pick-up fiber is placed in a specimen of mucosa or tumor; additional specimens of known thickness are stacked on top

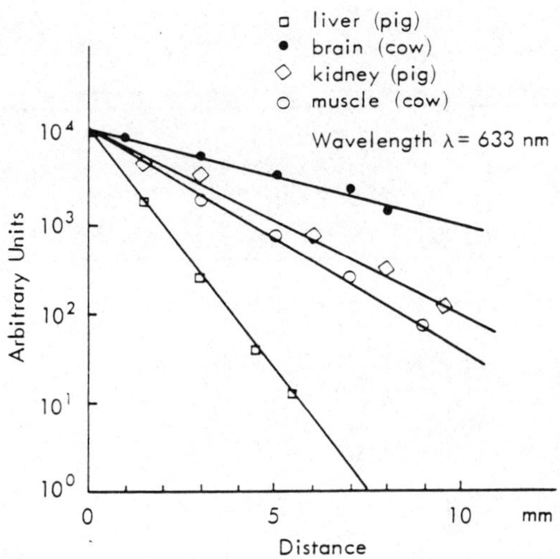

Figure 4. Space irradiance vs. depth penetration in various animal tissues <u>in vitro</u>. Tissues were from a commercial supplier and were subject to normal handling procedures for unfrozen meats.

additional specimens of known thickness are stacked on top to determine the attenuation coefficient. Embedding the fiber in the tissue eliminates edge effects.

Evaluation of these curves gives the optical attenuation coefficients in Table 4. The tumor tissue and the posterior portion of the bronchial membrane contained local areas of carbonaceous material. The material was black in color and could greatly limit the penetration of the light. The results given are based on selectively removing this carbonaceous material or by selection of specimens lacking this material.

Measurements using these techniques <u>in vivo</u> and <u>in vitro</u> on a variety of tissues at different wavelengths, are presently being obtained. Table 4 shows values obtained to date. The <u>in vivo</u>:<u>in vitro</u> comparisons show the relative effect of the loss of blood content in the tissues. The impact of blood, particularly hemoglobin, is most significant in the 400-600 nm region of the visible light spectrum due to its large absorption in this region (5). At 480-520 nm, both hemoglobin and oxy-hemoglobin show a secondary minimum in their absorption spectra; this explains the greater penetration in this region of the spectrum compared to results obtained at 545 and 577 nm. In the 630 nm region, hemoglobin absorption is minimal; this undoubtedly adds significantly to 630 nm light penetration in tissues.

Table 3. Attenuation coefficient, penetration depth and reflection coefficient for tissues *in vitro* (633 nm).

Tissue	Total Attenuation Coefficient $\alpha(cm^{-1})$	Penetration Depth $\delta=1/\alpha$ (mm)	Reflection Coefficient γ
Brain (Cow)	2.5	4.0	0.60
Kidney, Cortical Portion (Pig)	4.8	2.1	0.32
Muscle (Cow)	5.6	1.8	0.21
Liver (Pig)	13.0	0.8	0.18

Typically, the attenuation coefficient at 514.5 nm is approximately 2.0-2.5 times greater than at 630 nm. There is probably a significant blood loss from tissues under *in vitro* conditions. Penetration gains *in vitro* of 1.5 at 630 nm and 2.5 at 514.5 nm show the significance of this phenomenon.

Biological molecules in tissues such as melanin and other pigmentations also affect light penetration and will be the major factors dominating light penetration in dark tissues. The multiple values given for the cat brain at 630 nm and the rabbit muscle at 630 nm and 514.5 nm, represent measurements in different locations in the tissue. Variation in tissue makeup and consistency due to presence of fat and fluids, were found to correspond to areas of sizeable increase in penetration depth. Differences in the cow brain and cat brain are most likely due to the large content of blood in the latter. The cow brain used in the measurements was obtained from a commercial meat supplier and had been thoroughly flushed of blood.

POTENTIAL OF VARIOUS WAVELENGTHS OF LIGHT IN PRT

A number of investigators have raised the question of the the potential use of wavelengths other than red (630 nm) in PRT. The initial step in the photosensitizing reaction is the absorption of light by HPD component(s) localized in tissues. The efficiency of different wavelengths for generating the reaction would therefore be expected to follow the absorption spectrum of the component in the tissue. HPD exhibits the characteristic aetio-absorption of porphyrins

Table 4. Attenuation coefficient (α) and penetration depth (δ) of tissues in vivo.

TISSUE	λ (nm)	IN VIVO α (cm^{-1})	IN VIVO δ (mm)	IN VITRO α (cm^{-1})	IN VITRO δ (mm)
Cat Brain	630/632	5.0	2.0	5.3	1.9
		9.8	1.0	8.9	1.1
		4.4	2.3	-	-
		5.9	1.7	-	-
	577	25.9	0.39	-	-
	545	34.4	0.29	-	-
	514.5	-	-	13.3	0.75
	501.7	-	-	13.2	0.76
	496.5	-	-	13.2	0.76
	488	-	-	10.9	0.92
	405-410	44.1	0.23	46.1	0.22
Rat Sarcoma	630			2.3	4.4
	514.5			4.8	2.1
Muscle, Rabbit	630	1.6	6.1	1.1	8.8
		2.3	4.4	1.5	6.6
	514.5	4.8	2.1	2.0	5.1
		7.7	1.3	2.5	4.0
Brain, Cow (commercial)	633	-	-	2.5	4.0
Kidney, Pig	633	-	-	4.8	2.1
Muscle, Cow	633	-	-	5.6	1.8
Liver, Pig	633	-	-	13.0	0.8
Lung Tumor, Squamous Cell, Human	633	-	-	6.3	1.6
Bronchial Mucosa Membrane, Human	633	-	-	9.1	1.1
Lung Substance (deflated), Human	633	-	-	11.0	0.9

Table 5. Molar extinction coefficients of HPD and various forms of hemoglobin. In parentheses: ratio of ε_{HPD} to the ε_{Hbxx} for the Hemoglobin complex xx.

WAVELENGTH (nm)	Molar extinction coefficient (cm^{-1} M^{-1})			
	HPD†	HbCO‡	HbO_2‡	Hb‡
410	7.9+4	1.75+4 (4.51)	1.18+5 (0.67)	8.0+4 (0.99)
510	8.4+3	6.5+3 (1.29)	4.76+3 (1.76)	5.0+3 (1.68)
540	5.9+3	1.43+4 (0.43)	1.43+4 (0.41)	1.03+4 (0.57)
570	4.8+3	1.39+4 (0.35)	1.34+4 (0.36)	1.13+4 (0.42)
630	3.4+3	2.4+2 (14.2)	1.5+2 (16.0)	1.07+2 (31.8)

†reference 6
‡reference 5

with a major absorption band near 400 nm and four minor bands of decreasing magnitude at 500 nm, 540 nm, 570 nm and 630 nm (6). Studies by Kinsey et al. have shown this to be true *in vitro* (7). Effectiveness of a given wave length of light *in vivo* will depend on its absorption by the HPD localized component and also on competitive absorption by other materials. Comparison of the extinction coefficient of HPD in phosphate buffered saline (PBS) with 10% fetal calf serum (FCS), and extinction coefficients of the major forms of hemoglobin in blood are shown in Table 5. The ratio of the HPD coefficient to that of the hemoglobin complexes is given in parentheses. Based on these ratios, red light is the best choice for selective absorption by HPD. Green light (510 nm) is next best, with violet (410 nm), green-yellow (540 nm) and yellow (570 nm) progressively less effective.

Based on the light penetration data (Table 4), it is useful to make a theoretical comparison of the relative effectiveness, at depth in tissue, of the violet, green and red absorption bands of HPD. The relative photodynamic reaction at yield depth in tissue can be approximated using

Equation 1.

$$Y = \frac{\beta_{HPD}(\lambda)}{a_T(\lambda)} e^{-a_T(\lambda)x}$$

where:

Y = Relative Photoreaction Yield.

$\beta_{HPD}(\lambda)$ = (attenuation) absorption coefficient of HPD at wavelength (cm^{-1}).

= $2.3\,\varepsilon_{HPD}(\lambda)\,C$

$\varepsilon_{HPD}(\lambda)$ = HPD molar extinction coefficient (cm^{-1} M^{-1})

C = Molar concentration of HPD in tissue

$a_{Tiss}(\lambda)$ = Tissue attenuation coefficient (cm^{-1})

$a_T(\lambda)$ = Total attenuation coefficient of tissue at wavelength (cm^{-1}).

= $\beta_{HPD}(\lambda) + a_{Tiss}(\lambda)$.

x = Depth in tissue (cm).

equation 1 assuming absorption-dominant attenuation. This equation assumes that the relative photodynamic reaction yield (Y) is directly related to the amount of light absorbed by HPD, and that light transport is an absorption-dominant process. It does not consider physiological conditions which may determine biological effectiveness. Fig. 7 shows a comparison of 630, 510 and 410 nm light using Equation 1. These results are based on the assumption that localized HPD component(s) have an absorption spectrum comparable to that of HPD in PBS containing 10% FCS, and that the tissue attenuation coefficients are 2.0 cm^{-1}, 4.25 cm^{-1} and 45.0 cm^{-1} for red, green and violet light respectively. Note that in Equation 1, $a_T(\lambda)$ is the sum of the attenuation coefficient of the tissue and the HPD absorption coefficient, i.e., $a_T(\lambda) = \beta_{HPD}(\lambda) + a_{Tiss}(\lambda)$.

The HPD concentration is assumed to be 1.0 µg/gram tissue. Increasing the drug concentration moves the crossover point of the 510 nm and 630 nm light to smaller depths. At con-

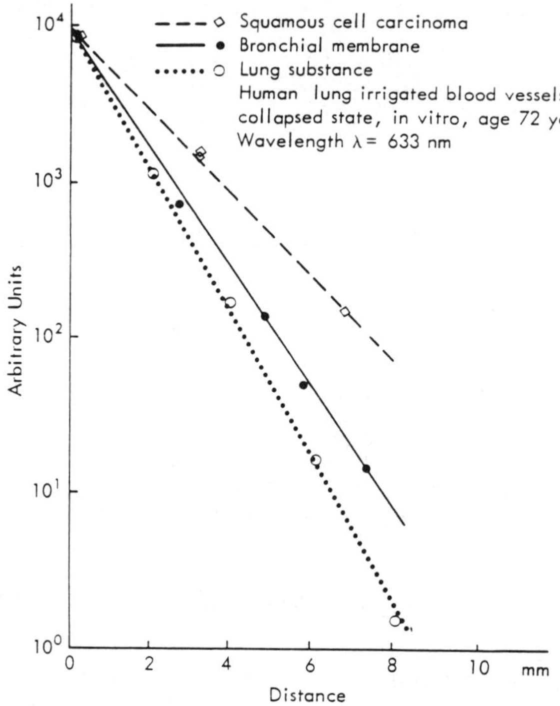

Figure 5. Radiance versus depth at 633 nm in human lung tissues (*in vitro*). Data is normalized to 0.0 nm.

centrations greater than 5 μg per gram of tissue, the HPD significantly adds to the total attenuation coefficient. This will lead to a decreased therapeutic dose at depth. For example, increasing the HPD concentration to 5 μg/gram, in the calculation for Fig. 7, increases the photoreactive yield by a factor of 2.4, instead of 5.0, at 1.0 cm depth for 630 nm light. At a 10 μg/gram concentration, the yield is only increased by a factor of 2.0 compared to the 1.0 μg drug concentration. This represents a net loss of 17% compared to the 5 μg drug level. This shielding effect of HPD in the tissue must be considered when treating large lesions at increased HPD dosages. Examination of Fig. 7 shows that at small thicknesses (<1.0 mm), 510 nm light would be more effective in producing a photodynamic reaction per incident photon. At depths > 1.0 mm, the 630nm light yield would be greatest and would be a factor of 10 times greater at 7 mm depth. The 410 nm light is so heavily attenuated by the tissue that its effectiveness would be only superficial. It must be stressed that this analysis is theoretical and needs to be evaluated *in vivo*. With the potential of delivering large quantities of 514.5 nm light from present CW argon lasers or 510 nm light from the recently-developed copper-

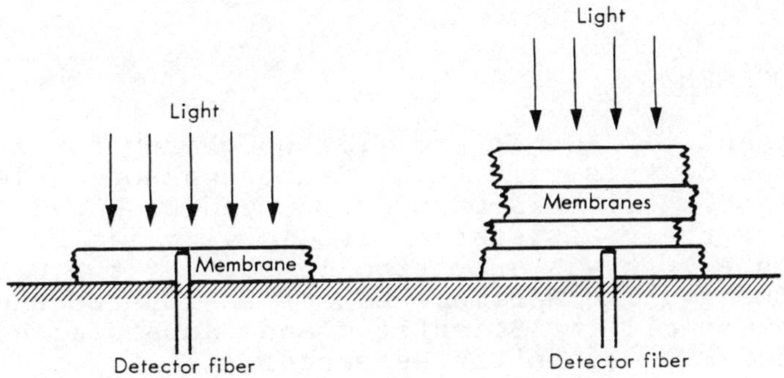

Figure 6. Schematic representation of method used to measure penetration depth in bronchial membrane and lung tumor specimens.

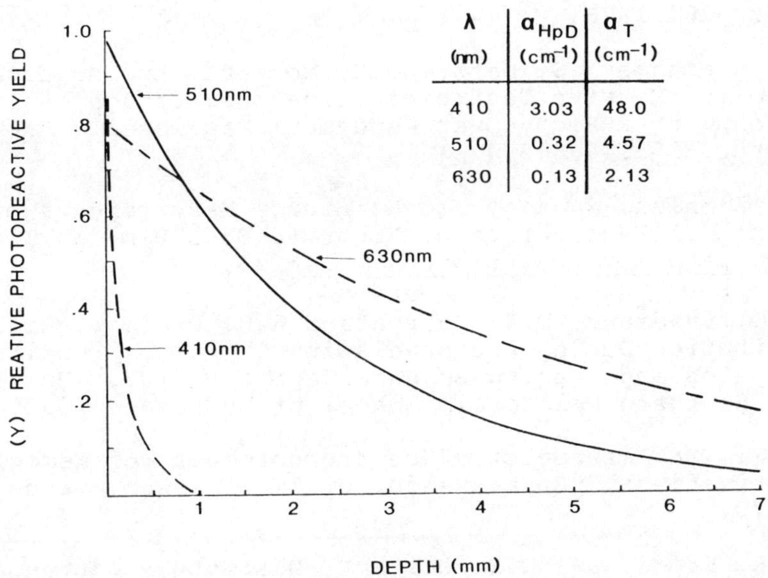

Figure 7. Theoretical photoreactive yield (Y) at depths for 410 nm, 510 nm and 630 nm light. Based on absorption properties of 1.0 μg/ml HPD in PBS with 10% FCS.

argon lasers, use of green light for superficial lesions may be promising, particularly when treatment at greater depth is not desired. Future work involving comparison of 514.5 nm vs. 630 nm light is planned.

ACKNOWLEDGEMENTS

We thank C.J. Gomer and T.J. Dougherty for stimulating discussions of this work, A.L. Castorena for assistance in the preparation of this manuscript, and D. Rounds and S. Jacques for their assistance with *in vivo* cat brain measurements. This work was supported in part by the U.S. Department of Energy, The National Cancer Institute, the Royal Norwegian Council for Scientific and Industrial Research and the Clayton Foundation for Research.

REFERENCES

1. K.R. Weishaupt, C.J. Gomer and T.J. Dougherty. Identification of Singlet Oxygen as the Cytotoxic Agent in Photo-inactivation of a Murine Tumor. Cancer Research, 36:2326, 1976.

2. P. Kiefhaber, G. Nath and K. Moritz. Endoscopical Control of Massive Gastrointestinal Hemorrhage by Irradiation with a High-Power Neodymium-Yag laser. Prog. Surg., 15:140, 1977.

3. J. Eichler, J. Knof and H. Lenz. Measurements on the Depth of Penetration of Light (0.35-1.0 m) in Tissue. Rad. and Environ. Biophys., 14:243, 1977.

4. L.O. Svaasand, D.R. Doiron and A.E. Profio. Light Distribution During Photoradiation Therapy. Institute for Physics and Imaging Science Report 900-02. University of Southern California School of Medicine, 1981.

5. O.W. van Assendelet. "Spectrophotometry of Hemoglobin Derivatives." Royal Vanforcum LTD., Assen, Netherlands, 1970.

6. A.E. Profio and D.R. Doiron. Dosimetry Considerations in Phototherapy. Med. Phys., 8(2):190, 1981.

7. J.H. Kinsey, D.A. Cortese, H.L. Moses, R.J. Ryan and E.L. Branum. Photodynamic Effect of Hematoporphyrin Derivative as a Function of Optical Spectrum and Incident Density. Cancer Res., 41:5020, 1981.

THERMAL DISTRIBUTION DURING PHOTORADIATION THERAPY

Lars O. Svaasand† and Daniel R. Doiron‡

†Division of Physical Electronics
University of Trondheim
Trondheim, Norway

‡Institute for Physics and Image Science
School of Medicine
University of Southern California
Los Angeles, California

INTRODUCTION

The optical power which is delivered into the tissue during photoradiation therapy will be absorbed and converted into heat. The resulting temperature rise may result in hyperthermal contributions to the cell kill or there may be synergistic effects between hyperthermal therapy and photoradiation therapy.

This paper discusses the extent of the regions with expected hyperthermal cell kill of malignant tumors in terms of tissue parameters and delivered optical dose.

Principles of thermal and optical distribution in tissue

The light distribution in most tissue is dominated by scattering. A collimated optical beam is found to be scattered into a rather isotropic distribution after propagating less than 1 mm in tissue. The ratio between the radiance in the forward and backward direction is typically less than one order of magnitude in 1 mm distance from a collimated source.

The propagation of light in the regions more than 1 mm

from the collimated source may thus be approximated by diffusion theory (1). The light distribution is now given by

$$\vec{J} \cdot -\zeta \operatorname{grad} \phi \tag{1}$$

$$\operatorname{div} \vec{J} = -\beta\phi + \kappa_S \psi \tag{2}$$

where $\vec{J}$ is the photon flux transport vector and ϕ is the space irradiance (8). The parameters ζ and β are the diffusion coefficient and the absorption coefficient, respectively. The unscattered light source intensity is ψ and κ_S is the scattering coefficient.

We shall, however, assume that the unscattered light is confined to an infinitely thin surface layer. The two equation may thus be expressed [Eqs. 1, 2]

$$\operatorname{div} \operatorname{grad} \psi - \frac{\beta}{\zeta} \phi = 0 \tag{3}$$

We shall first consider the case where the light is delivered into the tissue with constant intensity over a sphere. This is an idealization of the case where the light is delivered through an inserted optical fiber.

The equation for the space irradiance may for the centrally symmetrical case be expressed [Eq. 3]

$$\frac{1}{r} \frac{d^2}{dr^2} (r\phi) - \frac{\beta}{\zeta} \phi = 0 \tag{4}$$

The solution of this equation for the infinitely large medium can be written

$$\phi = \phi_o \left(\frac{a}{r}\right) e^{-\alpha(r-a)} \tag{5}$$

where ϕ_o is the space irradiance at position r = a. The total attenuation coefficient α is given by

$$\alpha = \frac{1}{\delta} = \sqrt{\frac{\beta}{\zeta}} \tag{6}$$

where δ is the optical penetration depth.

If we require continuity of the flow of optical power

at the interface between the light delivery system of radius a and the tissue, we have [Eq. 1]

$$\phi = \frac{P_t}{4\pi a\,(1+\alpha a)\zeta}\left(\frac{a}{r}\right)e^{-\alpha(r-a)} \tag{7}$$

where P_t is the transmitted power.

The absorbed optical power per unit volume q" may now be expressed [Eqs. 2, 7]

$$q'' = \beta\phi = \frac{P_t\alpha^2}{4\pi a(1+\alpha a)}\left(\frac{a}{r}\right)e^{-\alpha(r-a)} \tag{8}$$

The absorbed power is thus only dependent on the scattering coefficient and the absorption coefficient through the total absorption coefficient α.

The thermal distribution in tissue can be expressed (2,3)

$$\vec{J'} = -\kappa \text{ grad } T$$

$$\text{div } \vec{J'} = -\rho C\frac{\partial T}{\partial t} + q \tag{9}$$

where $\vec{J}$ is the heat flux vector, T is the temperature, is the thermal conductivity, ρ is the mass density, C is the specific heat per unit mass and t is the time. The term q represents the heat source or sink density.

The heat source density q is a vascular medium can be expressed in terms of two parts: the power density q' which is transported out of the unit volume by the capillary blood flow and the power density q" which is generated in the region through optical absorption. The first mechanism Q' may be represented by

$$q' = \rho CQ(T-T_o) \tag{10}$$

where ρ and C are the mass density and specific heat of the blood, respectively. These values are not significantly different from those of the tissue. The normal temperature of the tissue i.e. the arterial blood temperature is T_o. The blood flow is characterized by the blood flow Q measured in s^{-1}. The relation between this flow and the blood flow per unit mass of the tissue is

$$Q = \rho F \tag{11}$$

where the blood flow F is measured in m³/s kg (6.10 ml/min g).

The equation of heat in the *in vivo* case may thus be written [Eqs. 9, 10]

$$\text{div grad } T - \frac{1}{X}\frac{\partial T}{\partial t} - \frac{Q}{X}(T - T_o) = -\frac{q''}{\kappa} \tag{12}$$

where X is the thermal diffusivity given by

$$X = \frac{\kappa}{\rho C} \tag{13}$$

When the expression for the absorbed optical power density is substituted i the heat equation and the normal tissue temperature is normalized to zero, we have the steady state solution in the form

$$T = \frac{-P_t}{4\pi\kappa\left[1 - \left(\frac{a_V}{a}\right)^2\right](1 + aa)} \frac{1}{r} e^{-a(r-a)} + \frac{C_1}{r} e^{-a_V(r-a)} + \frac{C_2}{r} e^{+a_V(r-a)} \tag{14}$$

where a_V is the thermal attenuation coefficient given by

$$a_V = \frac{1}{\delta_V} = \sqrt{\frac{Q}{X}} \tag{15}$$

where δ_V is the thermal penetration depth (2). The temperature distribution is thus characterized by the optical penetration depth and the thermal penetration depth.

The optical penetration depth which is dependent upon pigmentation and scattering, is strongly wavelength dependent. The maximum value is found for red and near infrared wavelengths. The penetration depth in heavy pigmented tissue, as liver and lung, is of the order 1 mm for red wavelengths. The corresponding value for whiteish tissue, as brain is 4 mm. The optical penetration depth for red wavelengths is typically two to three times larger than the value for green wavelengths (1).

The thermal penetration depth is given by thermal diffusivity and blood perfusion. The thermal diffusivity is rather independent of the tissue. A typical value for muscle is 1.2×10^{-7} m^2/s (4). Typical value for the thermal penetration depth in tissue with high blood flow, as retina-choroid layer and cortex of kidney, is 1-2 mm. The penetration depth in tissue with intermediate blood flow, as brain and gastrointestinal tract, is 3-5 mm. The penetration depth in tissue with low blood flow, as ordinary muscle, is 7-9 mm. The mean thermal penetration depth in tumors is dependent upon tumor size. Reported value for blood flow in small tumors (0.3-0.7g) correspond to penetration depth of the order 4 mm. The corresponding value for larger tumors (2-5g) is 7 mm (5).

During photoradiation therapy the optical fiber is normally either inserted without any other protection than the plastic jacket, or the fiber is mounted into the lumen of a stainless steel hypodermic needle. The heat flow through the fiber itself or through the stainless steel tube is usually quite negligible.

If the dimensions of the tissue are much larger than the optical and thermal penetration depth, we may further consider the medium as being of infinitely large extent. When the constants C_1 and C_2 in Eq. (14) are determined from these boundary conditions, we have

$$T = \frac{P_t}{4\pi\kappa\left[1-\left(\frac{\alpha_v}{\alpha}\right)^2\right]} \frac{1}{r} \left\{ \frac{1}{1+\alpha_v a} e^{-\alpha_v(r-a)} - \frac{1}{1+\alpha a} e^{-\alpha(r-a)} \right\} \tag{16}$$

This expression is illustrated in Fig. 1 which shows the temperature rise per unit transmitted power vs distance from the center of the fiber. The optical penetration depth $\delta=1/\alpha$ is put equal to $\delta = 2$ mm. This value corresponds to the optical penetration depth for the wavelength region $\lambda =$ 610 - 640 nm in a squamous cell carcinoma of the human lung (1). The radius of the delivery system a is put equal to the typical fiber core diameter, i.e. 200 µm. The details of the actual optical distribution in the vicinity of the fiber end will usually differ from these idealized conditions. But since the light distribution is scattered to a more isotropic distribution after propagating less than 1 mm in most tissue the centrally symmetric approximation will be valid for larger distances. The finer details of the optical distribution in the vicinity of the fiber end will, however, only have a minor influence on the thermal distribution in distances larger than 1 mm. If the light is

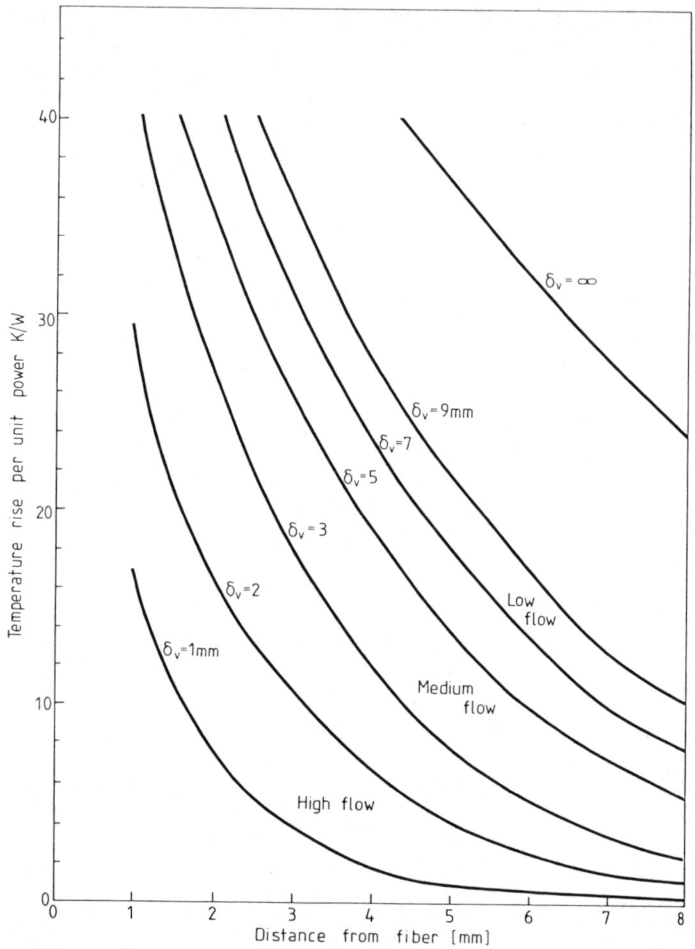

Figure 1. Temperature rise per unit optical power vs. distance as a function of blood flow. Optical penetration = 2 mm; thermal conductivity = 0.4 W/mK; radius of delivery system = 0.2 mm.

delivered into the tissue through the plane surface, we may, provided the linear dimensions of the illuminated region is much larger than the optical and thermal penetration depth, express the heat equation in the form of Eq. 12:

$$\frac{d^2T}{dx^2} - \frac{1}{X}\frac{\delta T}{\delta t} - \frac{Q}{X}(T-T_o) = -\frac{I t \alpha}{\kappa} e^{-\alpha x} \qquad (17)$$

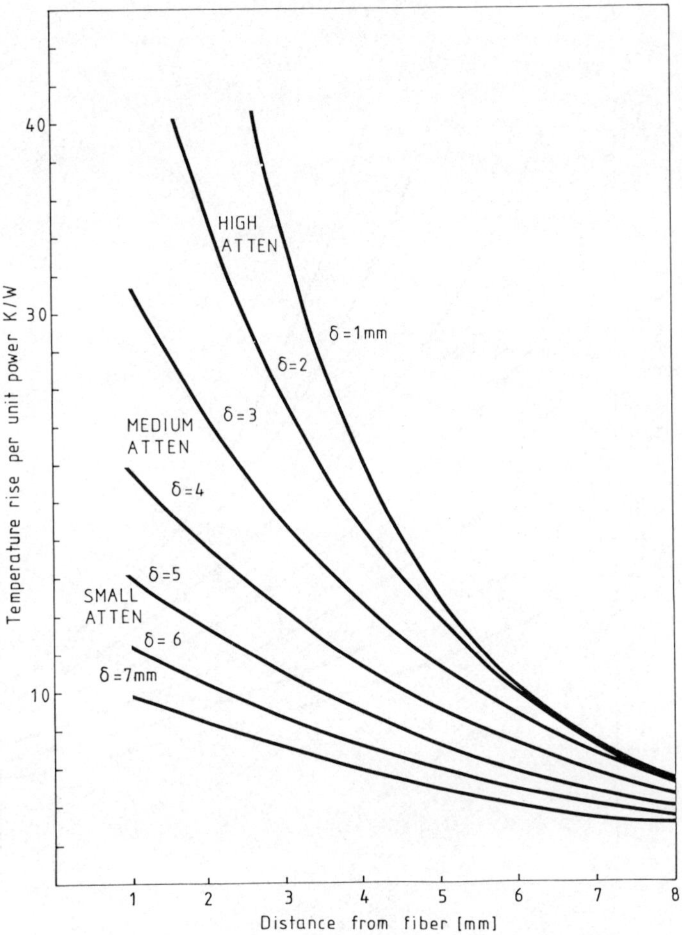

Figure 2. Temperature rise per unit transmitted optical power density vs. distance and blood flow. Optical penetration = 2 mm; thermal conductivity = 0.4 W/mK.

$$I_t = I(1-\gamma) \tag{18}$$

represents the transmitted optical power density and γ is the coefficient of reflection.

If the heat loss from the surface is negligible, we may put the solution in the form shown in Eq. 19.

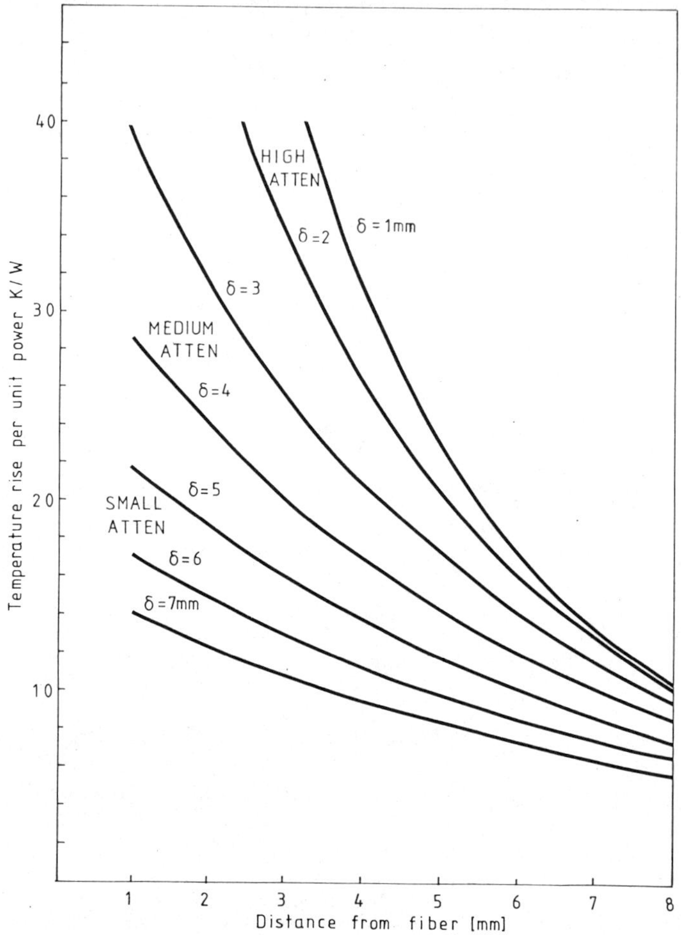

Figure 3. Temperature rise per unit optical power vs. distance and optical penetration depth. Thermal penetration depth = 2 mm; thermal conductivity = 0.4 W/mK; radius of delivery system = 0.2 mm.

$$T = \frac{a I_t}{\kappa(a^2 - a_v^2)} \left\{ \frac{a}{a_v} e^{-a_v x} - e^{-ax} \right\} \tag{19}$$

This equation is illustrated in Fig. 2 which shows the temperature rise per unit transmitted power density may be substantially lower than the incident. The coefficient of reflectivity at = 633 nm varies from $\gamma = 0.18$ in liver (pig) to $\gamma = 0.60$ in brain (cow) (1). Other parameters in Fig. 4 are the same as discussed in Fig. 1.

Hyperthermal contribution to cell kill

In therapy of tumors by local heating of the malignant tissue the temperature region from 41°C to 45°C is of particular interest. Most malignant cells are found to be more sensitive to heat than normal cells in this temperature region. This differential cell kill is partly lost for temperatures above 45°C, and all cells are killed rapidly for temperatures higher than 50°C.

Typical reported exposure times are 240 min at 41°C and 15 min at 45°C. An optimal thermal dose in local hyperthermia is to heat the margins of the tumor to 43°C for 60 min allowing the core to reach whatever higher temperature it achieves under the circumstances (6).

The exposure time which is necessary to introduce a given bioeffect (e.g. cell kill) is reported to be halved for each degree temperature rise above 42.5°C (6,7). The necessary exposure time Δ may thus be expressed

$$\Delta = E\, e^{-\xi (T_t - T_r)} \tag{20}$$

where T_t is the tissue temperature, the coefficient ξ is ξ = -ln 0.5 K^{-1}, the temperature T_r = 315.5 K (42.5°C), and the constant E is E = 5.2 x 10^3 s.

The temperature rise above normal tissue temperature in the actual temperature region will basically be proportional to the optical power. This temperature rise may be found from Eq. 16 for the case with an optical fiber inserted into the tissue and from Eq. 19 for the case of illumination through the surface.

The results are summarized in Tables 1 and 2. These tables visualize the extent of the region with expected hyperthermal cell kill in tumorous tissue with initial temperature T_O = 37.5°C. The effective exposure time is, due to the transient period, typically 5-10 min shorter than the total exposure time. Table 1 shows the case with an serted fiber and Table 2 shows the corresponding values for the case when the tissue is illuminated with a broad optical beam. Table 1 shows that an optical power of 400-600 mW may result in a hyperthermal cell kill in a tumor with size 0.5 g if a fiber is inserted in the center and the exposure time is of the order 20-30 min. Table 2 shows that a transmitted power density of 300 mW/cm² may result in hyperthermal cell kill in a depth of 5-10 mm into a subcutaneous tumor. The total power density must, dependent upon pigmentation of the skin, be up to twice as large.

Table 1. Extent of region with hyperthermal cell kill.

Transmitted optical power mW	Effective exposure time min	Minimum required temperature °C	Temperature rise per unit power K/W	Radius of region with hyperthermal bioeffect mm Blood flow		
				low	medium	high
600	60	43	9.2	7-8	5-6	2-3
400	60	43	13.8	6-7	4-5	1-2
200	60	43	27.5	3-4	2-3	0-1
600	30	44	10.8	7-8	4-6	2-3
400	30	44	16.3	5-6	3-4	1-2
200	30	44	32.5	3	2-3	0-1
600	15	45	12.5	6-7	4-5	1-3
400	15	45	18.8	5-6	3-4	1-2
200	15	45	37.5	2-3	1-2	0-1

Optical penetration depth δ = 2 mm
Radius of delivery system a = 0.2 mm
Fiber excitation

Table 2. Extent of region with hyperthermal cell kill. Broad beam excitation.

Transmitted optical power density mW/cm^2	Effective exposure time min	Minimum required temperature °C	Temperature rise per unit power density $10^{-2}K/mW/cm^2$	Extent of region with hyperthermal bioeffect mm Blood flow		
				Low	Medium	High
300	60	43	1.8	>10	5-10	1-2
100	60	43	5.5	>9	0-4	0
50	60	43	11	4-7	0	0
300	30	44	2.2	>10	4-10	0-1
100	30	44	6.5	>8	0-3	0
50	30	44	13	2-5	0	0
300	15	45	2.5	>10	4-9	0-1
100	15	45	75.	>7	0-2	0
50	15	45	15	0-4	0	0

Optical penetration depth δ = 2 mm

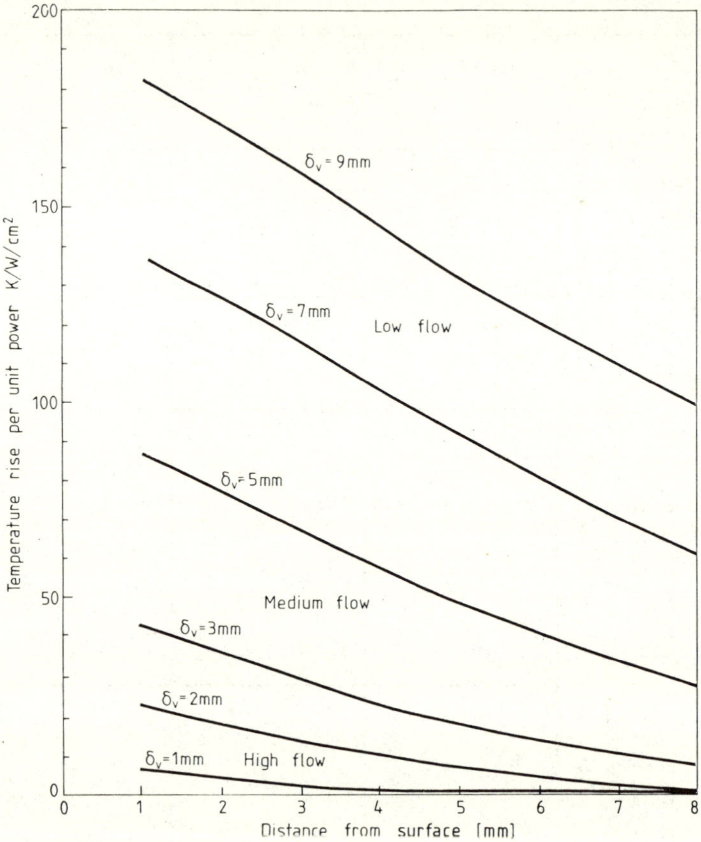

Figure 4. Temperature rise per unit optical power vs. distance and optical penetration depth. Thermal penetration depth = 9 mm; thermal conductivity = 0.4 W/mK; radius of delivery system = 0.2 mm.

The influence upon the thermal distributions from variations in the optical penetration depths are shown in Fig 3 (high blood flow) and 4 (low blood flow). flow). All curves show the temperature rise per unit transmitted power. The difference between optical power which is coupled into the fiber and the light which is transmitted into the tissue is usually quite small, since the specular reflected light in the interface between fiber and tissue is very small. Only a minor part of the scattered light will hit this interface.

CONCLUSIONS

The bioeffect in local hyperthermia is thus exponentially dependent on the optical power. This is in contrast to the photodynamic bioeffect in photoradiation therapy. This effect is usually anticipated to be proportional to the total number of photons in the relevant wavelength region, i.e. the delivered optical energy (8).

the necessary exposure time Δ in the case of photoradiation therapy may thus be expressed

$$\Delta = \frac{K}{P} \tag{21}$$

where K is a parameter dependent on optical properties of the tissue and the HPD concentration.

There are few detailed reported values for the parameter K for the *in vivo* case (9). The maximum reported values for optical power and exposure time during photoradiation therapy are in the region where the possibility for hyperthermal cell kill due to the temperature rise caused by the optical absorption must not be overlooked. The reported values for optical power in the low power limit indicate a minor temperature rise, and the bioeffect may be of an exclusively photodynamic nature.

Since the cell membrane is the primary target for both mechanisms, there may be synergistic effects for intermediate power levels. The necessary exposure time Δ in the case of interactions may be written [Eqs. 12, 20, 21]

$$\Delta^{-1} = \left(Fe^{-\xi \Lambda P}\right)^{-1} + \left(\frac{K}{P}\right)^{-1} + \left(g(P)\right)^{-1} \tag{22}$$

where g(P) is a parameter characterizing the interaction. The temperature rise is given by ΛP and F is a constant. The results indicate that the possibility for hyperthermal contribution or interaction must not be overlooked during photoradiation therapy. This is in particular important for the case where several hundred milliwatts of optical power is delivered through an optical fiber or for the case where several hundreds of milliwatts per square centimeters is radiated onto the skin.

ACKNOWLEDGEMENTS

The authors wish to thank T. Dougherty, C. Gomer and A.E. Profio for valuable discussions and comments.

The work has been carried out with the support from the U.S. Department of Energy, National Cancer Institute and the Royal Norwegian Council for Scientific and Industrial Research. The work has been carried out during one of the authors' (L.O.S.) period as visiting scientist at the University of California, Santa Barbara, Department of Chemical and Nuclear Engineering.

REFERENCES

1. Svaasand, L.O., Doiron, D.R. and Profio, A.E.: Light distribution in tissue during photoradiation therapy. Medical Imaging Science Group MISG, University of Southern California. USC-MISG 900-02 April 1980

2. Svaasand, L.O.: Propagation of thermal waves in vascular media. Clayton Ocular Oncology Center, Children's Hospital of Los Angeles, CHLA-COOP 100-01 Nov. 1980

3. Pennes, H.H.: Analyses of tissue and arterial blood temperatures in resting human forearm. J. Appl. Physiol. 1:93-122

4. Eichler, J., Knof, J., Lenz, H., Salk, J. and Schäfer, G.: Temperature distribution during laser irradiation. Biophysics, p. 277, Springer-Verlag 1978.

5. Song, C.W., Rhee, J.G. and Levitt, S.H.: Blood flow in normal tissues and tumors during Hyperthermia. J. Natl. Cancer Inst, 65: 1980.

6. Short, G.J. and Turner, P.F. Physical hyperthermia and cancer therapy. Proc. IEEE 68: 1, 1980.

7. Sapareto, S.A., Hopwood, L.E., Dewey, N.C., Rajer, M.R. and Gray, J.W. Effects of hyperthermia on survival and progression of Chinese hamster ovary cells. Cancer Res. 38: 393-400, 1978.

8. Profio, A.E. and Doiron, D.R. Dosimetry considerations in phototherapy. Med. Phys. 8(2), 1981.

9. Kinsey, J.H., Cortese, D.A., Moses, H.L., Ryan, R.J. and Branum, E.L. Hematoporphyrin derivative photodynamic effect as a function of optical spectrum and incident energy density. Proc. Mayo Clinic, 1981.

PHOTOBIOLOGIC CONSIDERATIONS IN PHOTORADIATION THERAPY

John A. Parrish

Department of Dermatology
Harvard Medical School
Massachusetts General Hospital
Boston, Mass. 02114

Photobiologic Considerations in Photoradiation Therapy

Hematoporphyrin derivative (HPD) photoradiation therapy of tumors depends on selective cell injury. The major selectivity of the treatment rests on an increased amount or prolonged retention of certain constituents or biotransformation products of HPD in tumor tissues. This increased concentration or prolonged retention results from some incompletely understood biochemical properties of malignant cells (protein binding, membrane changes, aberrant biotransformation) or cancer tissue (increased blood supply, avascular zones, absent lymphatics). Identification of the most active components of HPD and a better understanding of the mechanisms of prolonged retention may permit improved treatment design to achieve more selective destruction of tumor tissue.

Additional factors related to photobiology may increase selectivity. The mechanism of cell injury in photoradiation therapy is phototoxicity. In other studies, the term may have different significance. For example, in this report, phototoxicity is defined as injury or toxicity by photons with or without the use of exogenous photosensitizers. In the context of cutaneous photosensitivity, the term is often used to separate nonimmunologic photobiologic (phototoxic) responses from those effects involving immune mechanisms (photoallergic). Phototoxicity is sometimes used by clinicians to describe gross morphologic changes. It is also used by pathologists to describe histologic changes induced by ultraviolet radiation with or without photosensitizers regardless of molecular or cellular mechanisms. We suggest clearly defining phototoxicity before using the term.

On a microscopic and macroscopic scale one can select targets for a phototoxic response within tissue by selecting appropriate wavelengths likely to be transmitted to certain depths within tissue and likely to be absorbed by the appropriate chromophore. Other exposure dose parameters are available for manipulation and may improve selectivity of therapy. These include bandwidth (spectral power distribution) of exposure source, time of exposure, intensity of source and other parameters. This chapter will examine these variables.

Therapeutic Uses of Nonionizing Electromagnetic Radiation

Electromagnetic radiation results from transition of charged matter which converts either the potential energy held by separation of opposite charges, or from the kinetic energy associated with acceleration of a free charge, into transmittable energy. Photons interact with molecules within cells or tissue by the reverse process, imparting their energy into a displacement of charged matter which results in charge separation or acceleration. When a photon is thus absorbed, it ceases to exist. Central to photochemistry is the principle that only absorbed radiation can produce a chemical change. When a photon is absorbed, the excess energy is invested for some finite period of time in the atom or molecule, which is then said to be in an "excited" state. The mode of this excitation depends upon the amount of energy invested (photon energy). Photons with high energies, such as x-rays, may cause electrons to be removed from the absorbing molecule, resulting in ionization. Infrared photons (longer wavelengths, lower energy per photon) may alter lower-energy excitational levels by affecting rotational and vibrational states resulting in heat. Absorption of visible and ultraviolet radiation may lead to electronic excitation. Chemical changes may result. Cell metabolism may be altered or the cell may be killed.

Electromagnetic radiation in the ultraviolet and visible range is used with and without the addition of exogenous photosensitizers to treat a variety of diseases of skin, blood and other organic systems. Mechanisms of therapy vary (1). The therapeutic interactions of tissue and radiation from lasers is often explained by thermal effects. With conventional sources and some laser therapy the therapy depends on initial photochemical reactions within tissue. Photoradiation therapy is such an example.

In many skin disorders effectiveness of phototherapy is difficult to evaluate because the clinical improvement is unimpressive, the diseases are self-limiting, or controlled studies using bilateral comparison techniques have not been performed. Acne, atopic eczema, pityriasis rosea, parapsoriasis, and other diseases are occasionally improved by repeated exposures to sun or to artificial sources. In these dermatoses the therapeutic mechanism is not documented or understood.

In other instances therapeutic mechanisms of phototherapy are also not known with certainty but in vivo photochemical studies, assumptions about the pathophysiology of the disease processes, and knowledge of the nature of the interaction of electromagnetic radiation with biologic matter make it possible to list potential mechanisms (see Table 1). Phototoxicity is a likely mechanism for several forms of phototherapy and photochemotherapy. Paradoxically, certain photodermatoses can be improved by controlled exposures to ultraviolet or visible radiation. Phototherapy and photochemotherapy of photodermatoses may induce melanogenesis or injure cells involved in the pathogenesis of the photosensitivity. Depletion of mediators may also be involved. Phototherapy of psoriasis is moderately successful and widely used. It is assumed that photons must reach the proliferative compartment of the skin. DNA is thought to be an important therapeutic target because psoriasis is a hyperproliferative disease; UVR temporarily decreases DNA synthesis in normal skin; and other agents known to be effective in psoriasis, including tar, affect DNA.

Some forms of phototherapy appear to act by photochemical alteration of extracellular chemicals present in skin and blood. Generalized pruritus, a common symptom of chronic renal failure, has been reported to be diminished by 6 to 8 exposures of erythemogenic doses of UVB. In this situation it is possible that itching is caused by a photolabile metabolite normally cleared by the kidney. Intracutaneous extracellular photochemistry is the most likely mechanism in phototherapy of hyperbilirubinemia. While it is known that bilirubin can be photodegraded by photooxidative processes in vitro, this is probably not the only or most important therapeutic mechanism in vivo. Photoisomerization of unconjugated bilirubin results in a more water soluble form which can apparently pass directly into the bile.

Photochemotherapy is the use of a chemical plus subsequent nonionizing electromagnetic radiation to achieve a therapeutic benefit. An exogenous chromophore or photosensitizer is added to attempt to influence the site and nature of the photochemistry and subsequent photobiologic effects. Usually in the doses used the chemical alone or the radia-

Table 1. Phototherapy and Photochemotherapy

Possible Mechanisms:	Examples: Disease Treated
Phototoxicity of Abnormal Cells	Psoriasis Cancer (Hematoporphyrin Derivative) Polymorphous Light Eruption Mycosis Fungoides Lichen Planus
Induced Normal Protective Responses	Photodermatoses Vitiligo
Killing Organisms	Acne (+) Recurrent Herpes Simplex (-)
Alteration of Metabolite	Hyperbilirubinemia Uremic Pruritus Rickets

(+) No convincing evidence that disease process is altered.
(-) Double blind studies show no successful examples to date.

modified from: Parrish JA. Phototherapy and photochemotherapy of skin diseases. J. Invest. Dermatol. 1981; 77:167-171

tion alone have no significant biologic effect. The chemical may be given topically or parenterally but must reach viable cells which are subsequently exposed to radiation. There have been several examples of this form of *in vivo* photochemotherapy. The commonly used tar-UVB treatment of psoriasis was conceived and believed to be a form of phototherapy augmented by topical photosensitization with crude coal tar but this treatment as used today appears to be mainly UVB phototherapy. "Dye-light" treatment of recurrent herpes simplex with neutral red or proflavin and subsequent UVA or visible light has not proven to be an effective therapy. Psoralen-UVA (PUVA) therapy was discovered thousands of years ago and remains the best example of successful photochemotherapy.

Hematoporphyrin-visible light therapy of human cancer is also an example of photochemotherapy in the sense that an exogenous chemical is provided to cause an *in vivo* photochemical reaction. Hematoporphyrin and some of its derivatives absorb in the visible spectrum, which is an important waveband for all forms of photochemotherapy. The reasons are as follows (2):

1. This radiation is not directly very phototoxic. Human tissue other than the eye are not harmed by exposure doses most often used.

2. This radiation has photon energy which can cause electronic excitation in many organic molecules and therefore causes *selective* chemical reactions.

3. This radiation is transmitted into skin and cutaneous blood vessels and has an internal dosimetric depth profile; the penetration of 320-800 nm radiation into soft tissue is very wavelength-dependent because of the optical properties of the tissue.

4. This radiation is absorbed by a variety of molecules and bio-molecules which are nontoxic and can therefore be administered systemically.

Optical Properties of Tissue

The optical properties of human skin have recently been reexamined using new techniques (3) and models (4). The principle used for modeling dermis and blood can be applied to other organs. All photobiologic responses are influenced or determined by the optical properties of tissue, because these properties determine the dosimetry of optical radiation within the tissue.

When radiation strikes the skin or any other organ,

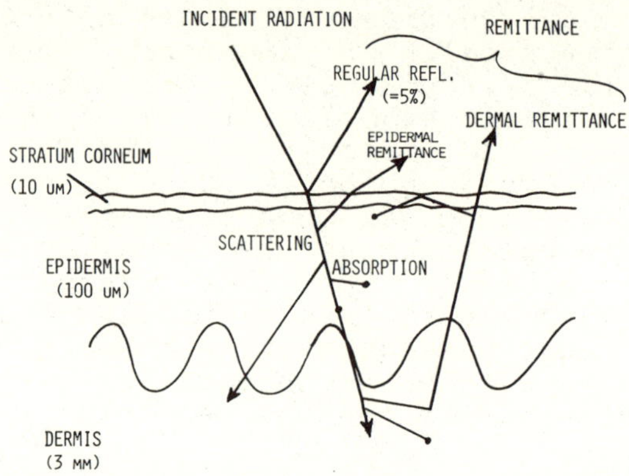

Figure 1. Diagrammatic representation of optical properties of skin.

part is scattered back to the environment, part is absorbed in various layers, and part is transmitted inward to successive layers of tissues, until the energy of the incident beam has been dissipated (Fig. 1). The first optical interaction with skin occurs at the surface of the stratum corneum, where some fraction of the incident radiation is scattered backward (reflected), since the refractive index of stratum corneum ($n_D \simeq 1.55$) is much higher than that of air (1.00). This fraction is called regular reflectance, and it is well described by Fresnel's law of reflection. The regular reflectance component is typically 5-7% for perpendicularly incident radiation, is nearly constant for all optical wavelengths, and is the same for normal black as for normal white skin (5). Because regular reflectance is relatively constant for all visible wavelengths, it does not in general greatly affect our perception of <u>color</u> of skin, but does account for the surface appearance of skin affecting impression of roughness, smoothness, glossiness, etc.

The term remittance is synonymous with <u>diffuse reflectance</u> and refers to the fraction of incident radiation which returns from the skin or from a sample, while <u>transmittance</u> refers to the fraction of incident radiation which passes through the skin or a sample. Remitted radiation includes the regular reflectance plus that radiation which enters the skin and is subsequently back-scattered outwards (Fig. 1). In the near UV, visible and near infrared regions, most of the remittance of Caucasian skin is, in fact, radiation which has traversed the epidermis, been back-scattered by the dermis, and returned again through epidermis, to the environment.

In the epidermis, aromatic amino acids both free and in

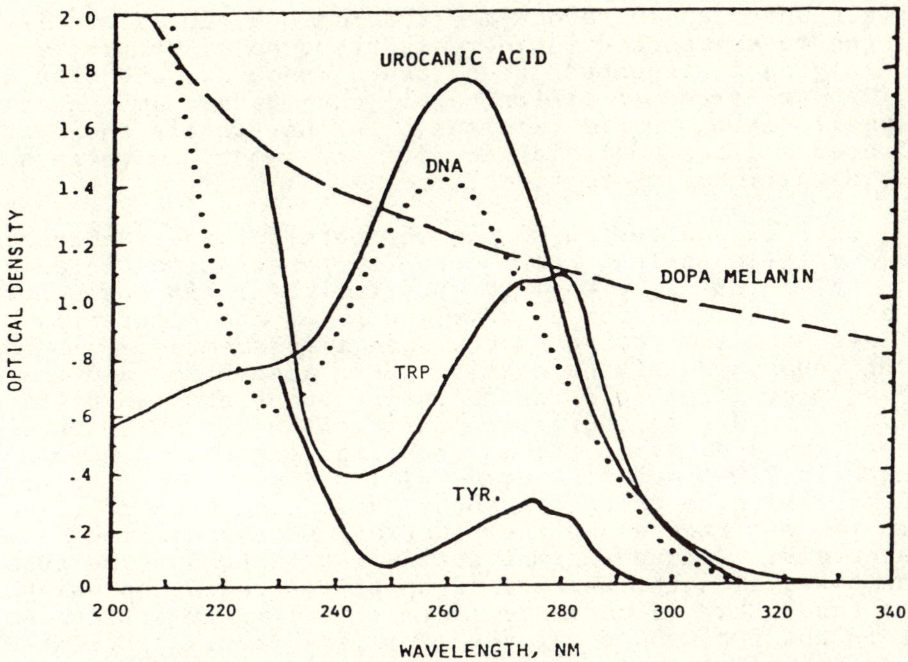

Figure 2. UV absorption spectra of major epidermal chromophores. DOPA-melanin, 1.5 mg% in H_2O; urocanic acid, 10^{-4} M in H_2O; calf thymus DNA, 10 mg% in H_2O (pH 4.5); tryptophan, 2 x 10^{-4} M (pH 7); tyrosine, 2 x 10^{-4} M (pH 7).

Modified from: Parrish JA. Responses of skin to visible and ultraviolet radiation. In: Goldsmith LA, ed. Biochemistry and pathophysiology of the skin. London: Oxford University Press, 1981: in press.

protein, urocanic acid, nucleic acids, and melanin are the major UV-absorbing chromophores (Fig. 2). Transmission of Caucasian stratum corneum or epidermis resembles that of a solution of protein containing the aromatic amino acids tryptophan and tyrosine and other aromatic compounds with a minimum in transmittance near 275 nm due to absorption by these chromophores. Nucleic acids, with an absorption maximum near 260 nm, and numerous small aromatic molecules, especially urocanic acid, with an absorption maximum at 277 nm at pH 7.4, also contribute to the broad 275 nm absorption band seen in the epidermis and stratum corneum. The absorbance of epidermis and stratum corneum for wavelengths less than 240 nm is largely due to peptide bonds. Melanin content and distribution also plays a major role in determining epidermal transmission of radiation. Across the near UV, visible and near infrared spectral regions, eumelanin is by far the most important epidermal chromophore, showing a smoothly declining absorption toward longer wavelengths out to 1200 nm, a result of its highly conjugated, amorphous polymeric chromophoric backbone. In the visible and near infrared regions, melanin can offer the most and perhaps only significant photoprotection.

Optical scattering within the dermis largely determines the average path-length and depth of penetration of radiation at various wavelengths. Scattering within the dermis is an inverse function of wavelength (6) and apparently derives from scattering by collagen fibers embedded in the ground substance of the dermis. Both absorption and scattering play major roles in radiation transfer within the dermis, to varying degrees in different wavebands. Optical absorption spectra of the major dermal pigments, and of synthetic "DOPA-melanin" are given in Fig. 3. At the wavelengths where hemoglobin absorbs, vasodilatation decreases both the remittance and the penetration of radiation into the dermis. Because dermal scattering is an inverse function of wavelength, the 50% or so of UVA radiation which penetrates fair Caucasian epidermis is largely attenuated within the first 50 μm of the papillary dermis, whereas 500 nm radiation may penetrate several hundred microns, and red and near infrared radiation several millimeters or more.

Of primary interest in considering photoradiation therapy is an understanding of the depths to which optical radiation penetrates. Despite the optical variables discussed above, the properties of human skin are constant enough (3) that a general picture emerges (Table 2). Regardless of melanin content, the 250 to 300 nm region is strongly absorbed by epidermis. The epidermal transmittance of 330 to 1300 nm radiation is high for fair Caucasian skin, but decreases markedly at progressively shorter wavelengths as melanin content is increased. In the dermis and in other organs, especially those with much connective tissue, optical scattering is an inverse function of wavelength and limits the penetration of shorter wavelengths. Blood absorbs strongly in the blue and green visible region. Between the infrared absorption bands of water and the green-visible absorption bands of hemoglobin lies the spectral region of greatest penetration.

Ultraviolet and visible radiation reaches the level of cutaneous blood vessels. Endothelial cells and connective tissue elements may be directly affected by the radiation; blood cells, lymphatics, and neural processes passing through the skin may be photochemically altered. Considering its metabolic needs, the skin is highly overperfused with average flow 20 to 30 times the minimal flow needed to supply skin cells. This is because cutaneous vasculature is

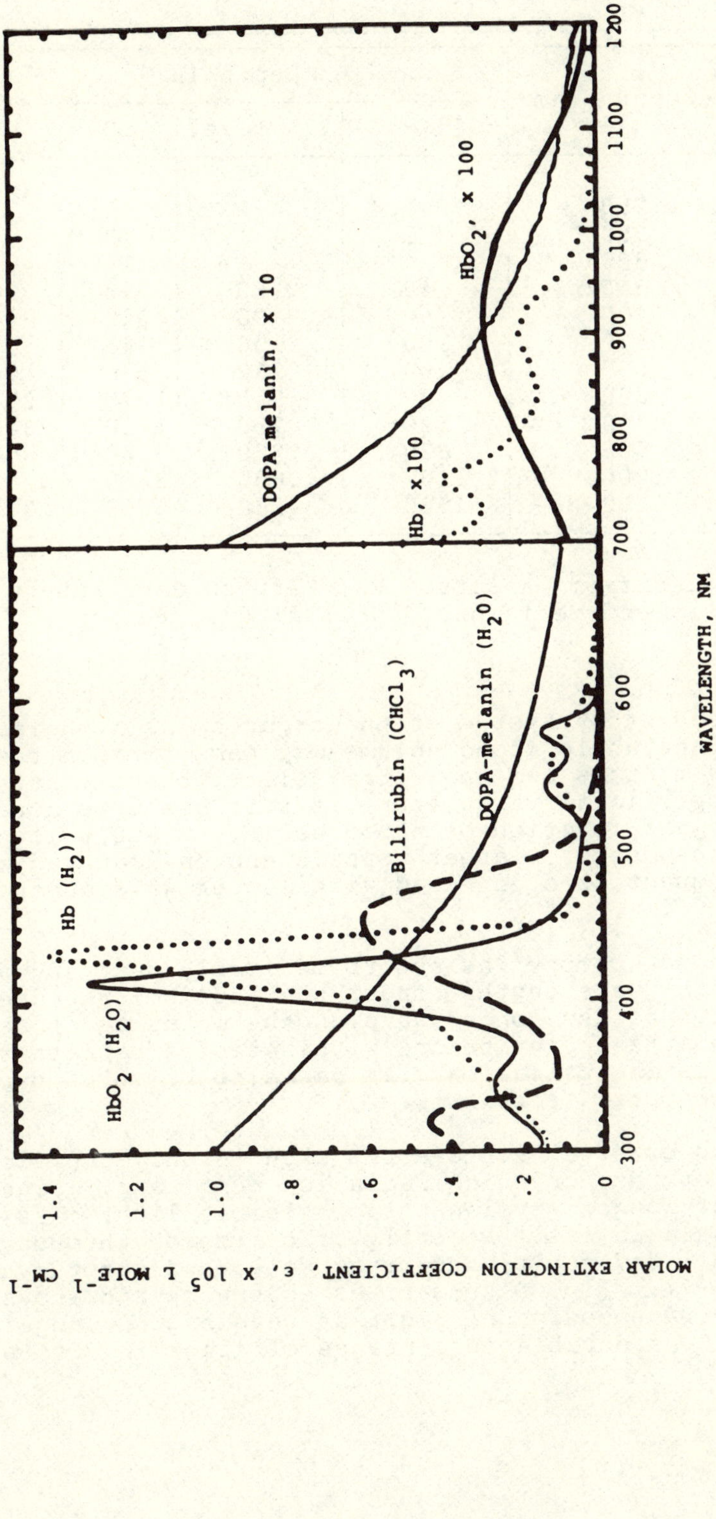

Figure 3. Important light-absorbing chromophores in human skin

Table 2. Approximate depth for penetration of optical radiation in fair Caucasian skin to various percentages of the incident energy density

Wavelength (nm)	Depth (μm)			
	50% (HVL)	37% (1/e)	10%	1%
250	1.4	2	4.6	9.2
280	1	1.5	3.5	7.0
300	4	6	14	28
350	40	60	140	280
400	60	90	200	400
450	100	150	345	690
500	160	230	530	1060
600	380	550	1270	2540
700	520	750	1730	3460
800	830	1200	2760	5520
1000	1100	1600	3680	7360
1200	1520	2200	5060	10120

modified from: Anderson RR, Parrish JA. Lasers in Surgery and Medicine. 1981; 1:263-276

a heat-regulatory system of the organism as a whole. Equivalent of the whole blood volume may pass through the skin every few minutes and the ultraviolet intensity at the level of the superficial vasculature is sufficient to alter the viability and function of blood cells. Visible light reaches the deeper, larger vessels and one can radiate significant quantities of blood with longer wavelength visible light.

Although porphyrins absorb maximally near 400 nm and much less at wavelengths longer than 550 nm, the deeper penetration of the longer wavelengths makes these the most useful for therapy of tumors within tissue beneath the skin. The skin as an optical barrier can also be often bypassed by using fiberoptic techniques.

Using external sources the skin is only the first optical barrier through which radiation must pass. Significant amounts of longer wavelengths of visible light pass through other tissue layers. Recently transmission through chest wall, skull and scalp, abdominal wall, and scrotum has been measured (7). Fig. 4 summarizes this work and shows significant transmission of light at wavelengths longer than 600 nm. Considerable percentages of light in 700-800 nm can

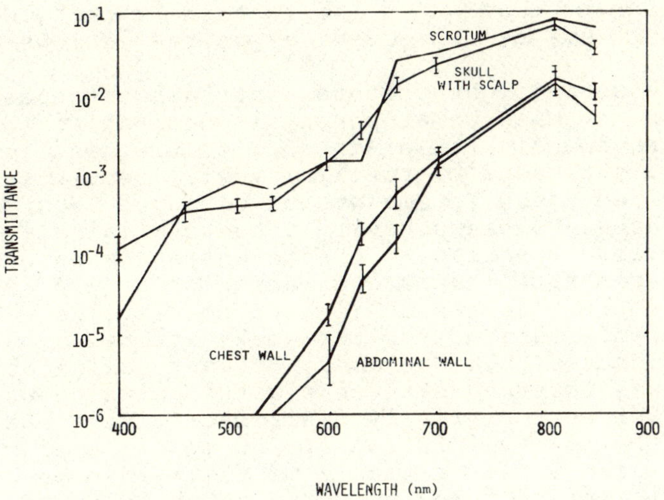

Figure 4. Spectral transmittance of various human tissues.
Range of thickness:
Skull with scalp: 9-13 mm (n=3)
Chest wall: 15-23 mm (n=5)
Abdominal wall: 22-32 (n=5)
Scrotum (two layers folded): 7 mm (n=2)
Shown at each point: mean ± standard error.

Modified from: Wan S, Parrish JA, Anderson RR, Madden M. Transmittance of nonionizing radiation in human tissues. Photochem. Photobiol. 1981; 34:679-681.

reach various human cavities.

Inflammation Resulting from Phototoxicity

Photochemistry within cells can lead to a host of chemical, metabolic and structural changes in both normal and abnormal cells. If enough cell injury occurs an inflammatory reaction follows. After a latent period of several hours, redness, swelling and tenderness are present in exposed sites. Of all cardinal signs of UV-induced phototoxicity of skin, the delayed erythema component has received the most attention because it is easily observed by noninvasive techniques. UV-induced delayed erythema is, however, difficult to quantify and its absence cannot be interpreted as absence of cell injury. Epidermal cell death (8), DNA damage (9), and melanogenesis (10,11) can be seen after suberythemogenic exposures to ultraviolet radiation. Recent studies have shown that suberythemogenic doses of UV alter the skin for at least 24 hr so that the threshold for delayed erythema by subsequent exposures is lowered (11).

HPD-visible light phototoxicity also induces cutaneous inflammation. When hematoporphyrin derivative (HPD) is infused intravenously for diagnostic or therapeutic purposes in humans, cutaneous phototoxicity will result in all persons provided adequate amounts of drug and light are used. In most cases photosensitivity appears clinically similar to a sunburn-like reaction. In some cases a more severe edematous reaction occurs and lasts for several days. In many cases the reaction could have been avoided if advice to remain out of sunlight for two weeks following infusion had not been ignored. In one case, however, clinical photosensitivity was observed six weeks after infusion of HPD. The photosensitivity originally described by Meyer Betz in himself (12) following infusion of 200 mg of hematoporphyrin also persisted for more than 2 months. In some patients in whom local extravasation occurred at the site of intravenous infusion, severe local photosensitivity occurred leading to a blistering reaction upon exposure to light.

In two patients in whom more severe phototoxic reactions occurred after the infusion of larger amounts (7 mg/kg) of hematoporphyrin derivative as part of an opthalmological evaluation, careful studies of time course and action spectrum for photosensitivity were performed (13). After an early (45 minute) peak of plasma hematoporphyrin levels, values fell to less than 5% of the peak within 3 days. A slower decline then occurred over the following two months, suggesting at least two pools of hematoporphyrin, perhaps with localization in the skin for extended periods. Repeated enterohepatic circulation could also be a factor.

Using a fluorescent UVA source filtered through window glass, a phototoxic response was elicited by 0.2 and 1.9 J/cm^2 UVA in the two patients 45 minutes after infusion. A decline in photosensitivity occurred in one patient over the first three days but the energy requirement for phototoxicity then remained at nearly the same level for two months. One patient was receiving β-carotene during the period and required approximately twice as much energy as the untreated patient to elicit phototoxicity. β-carotene may exert its effect by quenching excited singlet oxygen or free radicals formed by photon-induced reactions. Wavelengths between 400 and 450 nm were the most effective in causing cutaneous phototoxicity, with greater activity at 405 than at 365 or 425 nm.

The localization of HPD in skin has been studied in a limited fashion. In rats, HPD fluorescence in the skin 2 days after intradermal injection is mostly perivascular and in the keratin layer (14). Within the cell, hematoporphyrin accumulates in the soluble fraction (15), not in the cell wall, nuclei, mitochondria or microsomes of tumor cells. After administration of HPD to rats, irradiation leads to a rapid (15 minutes) increase in vascular permeability shown by trypan blue leakage and by carbon labeling of venules (14). Electron microscopic studies of fibroblasts cultures treated with hematoporphyrin and light (16) showed two patterns of damage, with either bleb-like cytoplasmic sacs appearing over the cell membrane and vacuolization of the cytoplasm, or a pattern of nuclear blurring and cytoplasmic vacuolization. Beta-carotene protected against both effects.

Increasing Selectivity

Selective killing of malignant cells in photoradiation therapy is primarily due to differences in tissue concentration of drug in normal and malignant tissues, and is also dependent on the number of photons absorbed for photochemistry. At the microscopic level this selective effect may be enhanced by choosing appropriate wavelengths to facilitate differential transmission and absorption within the tissue. As HPD is carried to tissue via blood, one could also manipulate blood flow by applying heat, cold, pressure or suction, or using pharmacological agents. Change in blood content may be used to alter the optical properties of tissue. If there is a difference in the oxygen dependence of the phototoxic effect on normal and abnormal cells, manipulation of blood flow may also make it possible to maximize killing of malignant cells and minimize inflammation or injury of normal cells.

Selectivity may be increased by thermal and photobiologic mechanisms. It is possible that different molecular mechanisms exist for the killing of normal and abnormal cells. It is likely that some of the photochemistry and molecular mechanisms involved in the killing of abnormal or malignant cells differ from those involved in induction of inflammation. The various effects of *in vivo* HPD photochemistry may result from different target sites (DNA, membrane, enzymes) within cells or from a difference in ability to quench, repair, reverse or compensate for toxic photoproducts or photon-induced lesions. Beneficial results can be increased and risks decreased by altering the way photons are delivered. Examples of manipulable treatment variables worthy of investigations include the following:

1. *Spectral power distribution of light source*. Action spectra represent effectiveness for a specified biological effect as a function of wavelength. It should be possible to compare precise action spectra for killing of normal and tumor cells and for induction of inflamma-

tion. If differences occur, especially at longer wavelengths, which penetrate more effectively, exposure sources could be tailored to maximize desired therapeutic responses. Tissue optics must be considered.

2. Intensity of light sources. The presence and degree of delayed erythema induced by exposure to ultraviolet radiation are dependent upon exposure dose. For a given area or tissue volume, the exposure dose equals the product of irradiance and exposure time. Within extremely wide limits (10^{10}-fold range) delayed erythema response is independent of irradiance ("dose rate"); that is, the influence of doubling the irradiance may be compensated for by a halving of the exposure time (17). This relationship, of a given exposure dose yielding a constant biologic response, is termed the reciprocity law. Most photobiologic responses that follow the reciprocity law over a wide range of irradiances may be assumed to derive from photochemical causes. That is, constant quantities of some photochemical reaction products are produced per absorbed photon at given wavelengths, and these products yield a constant biologic response. Thus, the delayed erythema response to UV exposure is initiated mainly via photochemical products. It is important to realize, however, that not all photochemically derived biologic responses necessarily follow the reciprocity law. The complexities of repair systems, inducible responses, factors affecting the transport, localization and stability of photochemical products, and the complicated cascade of events that ultimately lead to a response may obscure the fact that the primary photochemical event itself follows a reciprocity relationship. In addition, under saturating conditions in which a significant percentage of the absorbing molecules are excited at any given time, the response may not follow a reciprocity relationship. Reciprocity may hold over wide ranges of intensity for HPD-visible light-induced inflammation. Appropriate studies have not been done. However, preliminary studies in our laboratories (18) suggest that for HPD/visible light tumor killing intensity-time reciprocity does not hold. In one study we treated subcutaneous tumors in the flank of newborn rats. These flank tumors had been induced by injecting cells from FANFT[1] induced rat bladder tumors of syngeneic animals. Twenty-four hours after intraperitoneal injection of HPD (1.5 mg/100 grams) the same dose of 620 nm radiation (4 nm HBW) was delivered at different intensities (7.5, 28, and 75 mW/cm^2). This treatment resulted in extensive tumor cell destruction with little evidence of damage to surround-

[1]N-[4-(5-nitro-2-furyl)-2-thiazolyl]formamide.

ing skin. For a given dose, greater tumor necrosis occurred from the lower intensity irradiation given over longer exposure time. Therefore, under some conditions it should be possible to induce more tumor necrosis with less inflammation at longer exposure times and lower intensities. More detailed and thorough work is needed.

3. Waveband interactions and temperature. Infrared or visible radiation not initiating photochemistry may heat the tissue and increase competing or synergistic chemical or biologic reactions. Furthermore, visible light may activate or deactivate repair enzymes. The effects of temperature on photoradiation therapy deserve further studies.

4. Pulsed radiation. The use of very short pulses of light may lead to further spatial delineation of HPD-visible light-induced tissue necrosis, especially when multiple microembolic or micrometastatic disease is being treated, or when complex tumor-stroma patterns exist. It has been shown that with extremely short pulses of 577 nm dye laser radiation (<1 microsecond), microvasculature of human skin can be selectively damaged (19). This selective vascular effect results from a combination of selective absorption of radiation by blood and the use of a very short exposure time to confine thermal changes to a small volume of tissue. It may prove possible by similar approaches with pulsed lasers to selectively affect individual tumor cells or small tumor islands. The theoretical considerations involved in producing selective damage have recently been presented (20).

5. Selective protection. If different photochemical mechanisms and subsequent biochemical pathways do form a basis for selective phototoxicity, the use of β-carotene or other non-toxic chemical quenchers may diminish some manifestations of phototoxicity more than others, possibly resulting in increased selectivity of effect. Chemical or optical protectants may also be used to reduce unwanted effects. This could shift the risk/benefit ratio in favor of the patient.

CONCLUSIONS

Although some of the above photobiologic parameters can be tested *in vitro*, many of the tissue or organ responses such as tumor killing or inflammation can only be explored *in vivo*. The studies are often difficult, time consuming, and complex, and the resulting degree of increased selectivity in treatment may be small compared to that achieved by relative drug concentrations. Avoiding inflammation during therapy is certainly beneficial but only a part of the picture since patients may remain photosensitive for as long as 30 days after treatment. Despite these reservations, a better understanding of the photobiological factors listed above will make photoradiation treatment more rational and effective. Small selective advantages achieved by manipulating treatment conditions may be additive or synergistic. There is much to be done.

ACKNOWLEDGEMENTS

Dr. San Wan, Dr. Irene Kochevar, R. Rox Anderson and R. William Gange provided helpful criticism and suggestions.

REFERENCES

1. Parrish JA. Phototherapy and photochemotherapy of skin diseases. J. Invest. Dermatol. 1981; 77:167-171.

2. Parrish JA. New concepts in therapeutic photomedicine: photochemistry, optical targeting and the therapeutic window. J. Invest. Dermatol. 1981; 77:45-50.

3. Anderson RR, Parrish JA. The optics of human skin. J. Invest. Dermatol. 1981; 77:13-19.

4. Wan S, Anderson RR, Parrish JA. Analytical modeling for the optical properties of the skin with in vitro and in vivo applications. Photochem Photobiol. 1981; 34:493 493-499.

5. Anderson RR, Hu J, Parrish JA. "Optical radiation transfer in the human skin and application in in vivo remittance spectroscopy." In: Proceedings of the Symposium on Bioengineering and the Skin. Cardiff, Wales, July 19-21, 1979, MTP Press, Ltd., London. In press.

6. Anderson RR, Parrish JA. A survey of the acute effects of UV lasers on human and animal skin." In: Lasers in Photomedicine and Photobiology. Pratesi R, Sacchi CA, eds. pp 109-114. Springer-Verlag, Berlin, Heidelberg, New York.

7. Wan S, Parrish JA, Anderson RR, Madden M. Transmittance of nonionizing radiation in human tissues. Photochem. Photobiol. 1981; 34:679-681

8. Kaidbey KH, Grove GL, Kligman AM. The influence of longwave ultraviolet radiation on sunburn cell production by UVB. J. Invest. Dermatol. 1979; 73:343-345.

9. Gschnait F, Brenner W, Wolff K. Photoprotective effect of a psoralen-UV-A-induced tan. Arch. Dermatol. Res. 1978; 268:181-188.

10. Kaidbey KH, Kligman AM. The acute effects of longwave ultraviolet radiation on human skin. J. Invest. Dermatol. 1978; 72:253-256.

11. Parrish JA, Zaynoun S, Anderson RR. Cumulative effects of repeated subthreshold doses of ultraviolet radiation. J. Invest. Dermatol. 1981; 76:356-359.

12. Betz MF. Untersuchungen uber die biologisme (photodynamische) wirkung des hemotoprophyrins und anderer derivate des blut und gallenfarbstoffs. Dtsch. Arch. Klin Med. 1913; 112:476-503.

13. Zalar GL, Poh Fitzpatrick M, Krohn DL, Jacobs R, Harber LC. Induction of any photosensitization in man after parental exposure to hematoporphyrin. Arch. Dermatol. 1977; 113:1392-1397.

14. Barker DS, Henderson RW, Storey E. The in vivo localization of porphyrins. Br. J. Exp. Path. 1970; 51: 628-638

15. Winkelman J. Intracellular localization of hematoporphyrin in a transplanted tumor. J Nat Can Inst. 1961; 27:1364-1376.

16. Fritsch P, Gschnait F, Honigsmann H, Wolff K. Protective action of beta-carotene against lethal photosensitization of fibroblasts in vitro. Br J Dermatol. 1976; 94:263.

17. Anderson RR, Parrish JA. "A survey of the acute effects of UV lasers on human and animal skin." In: Pratesi R, Sachhi CA, eds. Lasers in photomedicine and photobiology. Proceedings of the European Physical Society. Quantum Electronics Division Conference, Florence, Italy, September 3-6, 1979. Berlin/Heidelberg/New York: Springer-Verlag, 1980; 109-114.

18. Anderson RR, Lin J, Lin C, Parrish JA. Unpublished data.

19. Greenwald J, Rowen S, Anderson RR, Harrist T, MacFarland F, Noe J, Parrish JA. Comparative histologic studies of the tunable dye (at 577 nm) laser and argon laser: the specific vascular effects of the dye laser. J. Invest. Dermatol. 1981; 77:305-310.

20. Anderson RR, Parrish JA. Microvasculature can be selec-

tively damaged using dye lasers: a basic theory and experimental evidence in human skin. _Lasers in Surgery and Medicine._ 1981; _1_:263-276.

PRECLINICAL EVALUATION OF HEMATOPORPHYRIN DERIVATIVE FOR THE TREATMENT OF INTRAOCULAR TUMORS: A PRELIMINARY REPORT

Charles J. Gomer*†‡, Bernard C. Szirth*‡, Daniel R. Doiron‡", James V. Jester*, Robert W. Lingua*, Corey Mark†‡, William F. Benedict†‡, and A. Linn Murphree*‡

*Division of Ophthalmology and
†Division of Hematology Oncology,
University of Southern California School of Medicine, Los Angeles, California

‡Clayton Center for Ocular Oncology[1]
Children's Hospital of Los Angeles
4650 Sunset Boulevard
Los Angeles, California 90027

INTRODUCTION

Treatment modalities such as external beam radiation therapy, cobalt plaques, photocoagulation or cryotherapy for preservation of useful vision in patients with ocular tumors offer only limited success. In many cases enucleation is necessary in order to prevent tumor spread (1). The accessibility of intraocular tumors to visible light and the excellent optical properties of the eye suggest that hematoporphyrin derivative photoradiation therapy (HPD PRT) may prove to be an important advancement in the successful management of intraocular and periocular malignancies. Due to the extreme sensitivity of the eye to light and the possible loss of vision resulting from non-repairable damage to normal ocular structures, we are performing preclinical studies designed to evaluate the efficacy of HPD PRT as a

---

[1]Address correspondence to Dr. Gomer at this address.

primary clinical modality for treating ocular tumors. Areas under investigation include; 1) development of relevant animal ocular tumor models; 2) quantifying HPD concentrations in normal and tumor tissue of the eye; 3) documentation of normal ocular tissue toxicity induced by HPD PRT and 4) examining the effectiveness of HPD PRT to selectively treat ocular tumors.

I. Animal Ocular Tumor Models

We are currently utilizing two experimental ocular tumor models to examine the effectiveness of ocular HPD PRT. Human retinoblastoma heterotransplanted to the anterior chamber of athymic (nude) mice has been utilized by our group to demonstrate tumor regression following HPD PRT (2) and to obtain quantitative levels of [^{3}H]-HPD in normal and malignant tissue (3). We have demonstrated that most of our retinoblastomas grown in the anterior chamber of the nude mouse maintain a similar karyotype and show the same degree of differentiation when compared to the primary tumor analyzed at the time of enucleation (4). The human retinoblastoma model is slow growing and therefore provides us with the opportunity to examine such parameters as HPD retention and localization. Unfortunately, it is not satisfactory for use as a model to adequately test ocular HPD PRT due to the extremely small size of the mouse eye, the anterior chamber placement of the retinoblastoma and the non-pigmented characteristic of the mouse.

Our second ocular tumor model is the amelanotic hamster melanoma (5). This lesion is routinely transplanted and grown in the anterior chamber of pigmented rabbit eyes. However, it is the ability to successfully implant this tumor in the rabbit choroid which makes this model important for HPD PRT studies (6,7). The availability of both anterior chamber and choroidal melanoma tumor models in the rabbit has enabled us to examine HPD distribution and HPD PRT induced tumor destruction. We are currently utilizing the anterior chamber tumor model to evaluate dose rate parameters, hyperthermal effects of HPD PRT and the effectiveness of various wavelengths of visible light in the HPD PRT process. The choroidal tumor model is currently being utilized to optimize our light delivery procedures prior to clinical treatment of intraocular tumors and to obtain additional HPD PRT dose response data.

II. HPD Distribution in Normal and Tumor Tissue of the Eye

Information regarding the localization characteristics of HPD in either ocular tumors or normal ocular tissues has been limited. Preferential localization of HPD has been observed in transplanted melanoma grown in the anterior chamber of hamsters and in transplanted VX-2 carcinoma grown in the anterior chamber of rabbits (8). HPD localization has also been demonstrated in the amelanotic melanoma transplanted to the rabbit choroid (7). Fluorescence was used as the sole indication of HPD presence, and in neither study was quantitative documentation obtained. We have recently determined that nude mouse eyes heterotransplanted with human retinoblastoma accumulated higher levels of [^{3}H]-HPD than control eyes (3). In addition, an improved therapeutic retention ratio of [^{3}H]-HPD (eyes with tumor compared to control eyes) was obtained at extended time intervals after administration (20 mg [^{3}H]-HPD/kg, 1-72 hr post injection).

With the use of the amelanotic melanoma transplanted in the rabbit anterior chamber, we have been able to obtain quantitative [^{3}H]-HPD levels for both tumor tissue and normal ocular tissue. Our study indicates that at 24 and 48 hr after intravenous administration of [^{3}H]-HPD (5.0 mg/kg) the tumor tissue retains 3.4 and 1.5 mg [^{3}H]-HPD/gm respectively. These levels were higher than any other ocular structure. The combined choroid-retinal pigment epithelium retains the highest level of [^{3}H]-HPD of all normal ocular tissues and this was approximately 60% of the level found in the tumor tissue. It should be noted that the concentration of [^{3}H]-HPD in the tumor tissue was significantly reduced at 48 hours after administration compared to 24 hours. This may have been a result of the rapid growth rate of this tumor. [^{3}H]-HPD levels in the cornea, lens, vitreous, retina and sclera were almost undetectable using our radioactive assay system (less than 0.25 μg/gm).

III. Normal Ocular Tissue Toxicity Induced by HPD PRT

Thorough preclinical toxicological studies are mandatory prior to any extensive clinical application of HPD PRT for the treatment of ocular tumors. We have performed an acute toxicity study utilizing 41 pigmented rabbit eyes and a chronic toxicity study has been started. Our acute toxicity program was designed to provide a dose-response relationship for damage to normal ocular tissue exposed to HPD PRT. At 48 hours following an intravenous injection of HPD (1 - 10 mg/kg), a 1.0 cm^2 area of the retina was exposed for 15 minutes to varying intensities (40-400 mW/cm^2) of red light (630 nm) generated by a rhodamine-B dye laser. Assessment of damage following treatment was performed using four parameters; documented visual examination by fundus photography, electroretinography, fluorescein angiography and histological examination by light microscopy. Damage was limited to the treatment field in all but the highest dose of HPD PRT. Fundus examinations demonstrated that toxicity in the forms of retinal edema, detachment and hemorrhage could be observed as early as two days following a single PRT treatment (5.0 mg HPD/kg, 100 mW/cm^2). Minimum doses of HPD PRT which induced visible or histological damage to the normal retina included 2.5 mg HPD/kg and 100 mW/cm² or 10 mg HPD/kg and 40 mW/cm². In addition, damage in the form of retinal necrosis was observable 2 days following a 15 minute exposure to red light at 400 mW/cm² at 400 mW/cm² in the absence of HPD.

Electroretinography (ERG) was performed on both control and treated eyes of rabbits prior to HPD PRT and then again 4 days and 13 days post treatment. We observed no changes in ERG recordings of treated eyes when compared to control eyes. These results indicate that the functional status of the large majority of photoreceptor cells and inner retinal cells were normal even though portions of the retina were obviously damaged following high doses of HPD PRT.

Fluorescein angiography is a procedure used in ophthalmology to evaluate the integrity of the vasculature of the eye. We performed fluorescein angiography on all treated rabbit eyes 10 days following HPD PRT. Vessel occlusion in the medullary ray was observed in rabbits treated with HPD PRT dosages consisting of at least 2.5 mg HPD/kg (with a 15 minute light exposure at 100 mW/cm²) or a 10 mg HPD/kg (with a 15 minute light exposure at 40 mW/cm²).

All animals entered in the acute toxicity study were sacrificed 14 days after treatment and the eyes were enucleated and sectioned. The thermal effects of light alone (400 mW/cm², 15 minute exposure, 630 nm, no HPD) included a well demarcated chorioretinal adhesion with loss of the outer layers of the retina and proliferation of the retinal pigment epithelium. An HPD dose of 10 mg/kg followed 2 days later by a 15 minute ocular exposure to 630 nm light at 100 mW/cm² resulted in massive retinal necrosis which extended outside the treatment area. At the other end of the response scale, an HPD dose of 5 mg/kg coupled with a 15 minute light exposure at a dose rate of 40 mW/cm² resulted in minor changes in the retinal pigment epithelium and retinal outer segments.

IV. HPD PRT of the Amelanotic Greene Melanoma

We are currently examining the effectiveness of HPD PRT in treating single nodule tumors (4.0 - 8.0 mm in diameter) growing in the anterior chamber of pigmented rabbit eyes. The treatment protocol is identical to that which we have used in documenting acute ocular tissue toxicity (described above). A direct comparison between normal ocular tissue toxicity and tumor damage can thereby be obtained. All animals are sacrificed 24 hours after treatment and HPD PRT induced tumor destruction is determined from histological sections obtained from the enucleated eyes.

Preliminary results obtained from 14 treated eyes indicate that the HPD PRT dose response for tumor treatment is similar to that which is observed in our normal ocular tissue toxicity studies. Dosages which are associated with significant normal retinal tissue toxicity (5-10 mg HPD/kg, 100 mW/cm^2) also induced massive tumor destruction. Dosages which gave rise to only minimal normal tissue toxicity (1-5 mg HPD/kg, 40 mW/cm^2) induced minimal tumor destruction. Additional experiments are in progress and must be analyzed in order to determine the definitive dose-response for tumor kill.

ACKNOWLEDGEMENTS

We thank Albert L. Castorena for assistance in the preparation of this manuscript. This work was performed in conjunction with the Clayton Foundation for Research and was supported in part by Biomedical Research Support Grant S07RR05469-17 from NIH.

REFERENCES

1. D.J. Apple and F.C. Blodi, Uveal Melanocytic Tumors - A Grouping According to Phases of Growth and Prognosis with Comments on Current Theories of Nonenucleation Treatment, in "Advances Regarding the Pathogenesis and Treatment of Ocular Tumors", D.M. Albert and R.R. Weichselbaum, eds, pp. 33, Little, Brown and Company, 1980.

2. W.F. Benedict, R.W. Lingua, D.R. Doiron, J.A. Dawson and A.L. Murphree, Tumor Regression of Human Retinoblastoma in the Nude Mouse Following Photoradiation Therapy; A Preliminary Report, Med. Pediatr. Oncol. 8:397 (1980).

3. C.J. Gomer, N. Rucker, C. Mark, W.F. Benedict and A.L. Murphree, Tissue Distribution of [^{3}H]-Hematoporphyrin Derivative in Athymic Nude Mice Heterotransplanted with Human Retinoblastoma, Invest. Ophthal. Vis. Sci. 22: 118 (1982).

4. W.F. Benedict, J.A. Dawson, A. Banerjee and A.L. Murphree, The Nude Mouse Model for Human Retinoblastoma: A System for Evaluation of Retinoblastoma Therapy, Med. Pediatr. Oncol. 8:391 (1980).

5. H.S.N. Greene and E.K. Harvey, The Growth and Metastasis of Amelanotic Melanomas in Heterologous Hosts, Cancer Res. 26:706 (1966).

6. D.L. Krohn, R. Brandt, D.A. Morris and A.S. Keston, Subchoroidal Transplantation of Experimental Malignant

Melanoma, Am. J. Ophthalmol. 70:753 (1970).

7. D.L. Krohn, R. Jacobs and D.A. Morris, Diagnosis of Model Choroidal malignant Melanoma by Hematoporphyrin Derivative Fluorescence in Rabbits, Invest. Ophthalmol. 13:244 (1974).

8. R.D. Cunningham and J.W. Henderson, Experimental Evaluation of Hematoporphyrin in the Detection and Management of Intraocular Tumors, Am. J. Ophthalmol. 6136 (1966).

PORPHYRIN LOCALIZING PHENOMENA

David Kessel and Ta-Hsu Chou

Departments of Pharmacology and Oncology
Wayne State University School of Medicine
Detroit, MI 48201

INTRODUCTION

The use of the porphyrin product HPD[1] for tumor localization (1,2) and therapy (2-5) has been described. One report on an analysis of HPD appeared in 1980 (6) and other such reports are contained in this book. In order to characterize the events leading to tumor localization by HPD, it will be necessary to delineate active components in the product, the nature of any porphyrin transformations associated with the localization process and loci of binding of localized fluorescent porphyrins within the tumor cell.

Some of our earlier studies were designed to utilize biophysical techniques for the study of cell-porphyrin interactions (7-11). Results of this work indicated that porphyrin uptake, the first step in photosensitization, was a function of drug hydrophobicity (7,11), that photodamage which occurred upon irradiation shortly after exposure of tumor cells to porphyrins involved the cell membrane (7), and that fluorescence and absorption spectra could be used to characterize loci of drug binding (9,10).

---

[1]Abbreviations: HPD, hematoporphyrin derivative [product of alkaline hydrolysis after acetylation]; HP, hematoporphyrin; HVD, monohydroxyethyl-monovinyldeuteroporphyrin; PP, protoporphyrin.

A report from Jori's laboratory (12) suggested that the maximal rate of singlet oxygen-mediated photooxidation occurred under conditions promoting drug dimer formation. We therefore examined spectral and photo-oxidative properties of different porphyrins to aid in characterizing both aggregation status and environment of intracellular porphyrins (9). Spectral studies can be useful for probing loci and aggregation status of intracellular porphyrins, but the resulting signals reflect a multiplicity of factors including the environmental dielectric constant, viscosity, and porphyrin aggregation status. To properly interpret spectra of localized porphyrin, the nature of the material must also be known. The present work describes our efforts in this regard. These result in the advancing of a hypothesis to account for localizing phenomena.

MATERIALS AND METHODS

Drugs

Hematoporphyrin (HP), hydroxyethyl,monovinyldeuteroporphyrin (HVD) and protoporphyrin (PP) were purchased from Porphyrin Products, Logan Utah. HPD was prepared from HP as described in Ref. 13 with minor modifications (11) to facilitate synthesis of small quantities of product. Solutions of porphyrins were made up in 0.1 M NaOH, and then brought to pH 7 with 0.1 M HCl.

For some studies, HP was purified by an extraction procedure. A 10 mg/ml aqueous solution of HP dihydrochloride was mixed with 2 volumes of methanol-chloroform (1:1). After shaking for 5 min, the phases were separated, the upper phase retained and extracted 20 times with fresh lower phase of a 1:1:1 water-chloroform-methanol mixture. The upper phase was evaporated under reduced pressure and the porphyrin dissolved in water. This extraction procedure removes PP, and diminishes by 20-fold the level of HVD, in the upper phase. The HP concentration was determined by measurement of optical density (399 nm) in 50 mM HEPES buffer containing 10 mM CTAB. With a 1 cm light-path, a 1 μg/ml solution of HP has an optical density of 0.180.

Fluorescence and absorption spectra

Spectral studies of porphyrins in different solvent systems were carried out using a dual-beam Perkin-Elmer 552 instrument with the light path varying from 1-10 mm. Fluorescence emission and excitation spectra were measured with a Perkin-Elmer MPF-44B instrument with appropriate filters to minimize light-scattering.

Biologic Studies

For studies *in vitro* we used the mouse sarcoma-180 cell line grown in RPMI 1640 medium + 10% horse serum and the murine lymphoblastic leukemia L1210 grown in MEM Eagle's medium + 10% horse serum. *In vivo* tumor localization was studied using mice sub-cutaneously implanted with the solid S-180 tumor. Animals were treated with porphyrins (7.5 mg/kg) 5 days after implant of 1 million cells; tumors were removed 2 days later.

Photosensitization studies

Using cells grown *in vitro*, photodynamic loss of cell viability was compared with inhibition of transport of the non-metabolized amino acid cycloleucine. Methodology described in Ref. 7 was employed:

1. Cells were loaded (10 min, 37°) with porphyrins in growth medium containing 10% serum, or in serum-free medium.

2. Drug-loaded cells were suspended in serum-free medium at 0°, to prevent any drug loss, and irradiated (550-650 nm fluorescent light, 1 mw/cm^2) for 10 min.

3. Irradiated cells were resuspended in complete growth medium for measurement of concentrative amino acid uptake using 5 min incubations at 37°.

4. Aliquots of control and treated cells were transferred to soft agar and colonies counted after 7 days as a measure of cell viability.

Drug uptake in vitro

Uptake of porphyrins by tumor cells *in vitro* was measured after incubations of 30 min or 24 hr at 37°. Cells were washed for 30 min in fresh growth medium (containing 10% serum) to remove readily-diffusible porphyrin. The relative intracellular level of different porphyrins was estimated by reverse-phase TLC studies on cell extracts.

Porphyrin analyses

Cells loaded with porphyrins *in vitro* or S-180 tumors removed from animals 2 days following porphyrin administration were dispersed in 50 mM HEPES buffer with a Polytron ultrasonic probe [100-300 mg (wet weight) per 2.5 ml]. The homogenate was centrifuged at 10,000 x g for 10 min, the top lipid layer discarded, and the remainder thoroughly mixed with 0.1 volume of 100 mM CTAB, heated to 85° for 3 min, cooled and shaken with 2 volumes of 1:1 chloroform-methanol. The lower phase containing the fluorescent material was concentrated *in vacuo* and the residue taken up in 1 ml $NaHCO_3$-washed ethyl acetate. The porphyrin was extracted into 1 ml of 50 mM sodium phosphate buffer pH 8.0, and this was eluted through a 0.5 x 1 cm reverse-phase column (BondElut, Analytichem, Harbor City, California). The column was washed with water, 50% methanol and 100% chloroform and the porphyrin eluted with 100% methanol, concentrated under nitrogen and used for TLC analysis. A slightly different extraction procedure is described in Ref. 14.

TLC studies were carried out using reverse-phase Whatman RP-18 plates, solvent = 65% methanol:35% 3 mM t-butylammonium phosphate pH 3.5. Standards were run with each analytic separation. The plates were dried, sprayed with kerosene to enhance fluorescence, and scanned (emission 395-400 nm; excitation 625-630 nm) using the TLC-scanning accessory of an Aminco fluorometer.

Fluorescence studies

Cells were dispersed in 50 mM HEPES pH 7.0 + 10 mM CTAB. Porphyrin content of cell homogenates was estimated by fluorescence, using broad band excitation (395-405 nm) and integrating the area under the fluorescence emission curve (600-720 nm). Fluorescence intensities were related to porphyrin concentration (expressed in terms of μg drug per 100 mg of cells, wet weight) by use of appropriate standards. These standards were also used to determine optimal excitation and emission wavelengths for each measurement.

Liposomes

To prepare HP-loaded liposomes, a mixture of 0.5 μmoles of stearylamine, 4.5 μmoles of dipalmitoyl phosphatidylcholine and 5 μmoles of cholesterol in 1:1 methanol-chloroform was evaporated (15) and taken up in 5 ml of HBS (140 mM NaCl + 10 mM HEPES pH 7.0) containing 0.5 mg/ml of purified hematoporphyrin. The mixture was shaken for 5 min under nitrogen at 50°, and then sonicated for 60 min at 37°. Optical clearing was noted after 6-10 min of sonication. The liposome-encapsulated hematoporphyrin was separated from free drug on a 2 x 40 cm column of Biogel P-2 using HBS as the eluting solvent. The first-eluted fluorescent fraction was retained. Analysis of liposomal porphyrin was carried out by bringing the pH of the suspension to 3.5 with HCl and extracting with 2 volumes of ethyl acetate. The solvent was removed under reduced pressure, and the porphyrins taken up in methanol for the reverse-phase TLC assay.

L1210 cells (20 mg) were incubated with 1 ml of HP-liposomes in HBS for 30 min at 37°. The cells were then collected by centrifugation (500 x g, 3 min). This centrifugation procedure will not sediment liposomes. The drug-

Table 1. Effects of Porphyrins on L1210 Cells

Porphyrin	Concentration (μg/ml)	CL transport % Control*	Viability†
HPD	6	51±6	210±38
HP(crude)	4.0	43±6	188±29
HP(purified)	60	54±5	210±31
HVD	1.0	52±3	225±17
PP	0.2	47±5	198±32

* % Control uptake of cycloleucine during 10 min incubations at 37° by cells previously loaded from medium containing specified drug levels and irradiated. Data represent mean ± S.D. of 5 trials.

† Number of colonies (control = 997 ±60) representing viable cells remaining after irradiation of drug-loaded cells. Values represent mean ± S.D. of 3 determinations.

loaded cells were suspended at 0° in HBS for 10 min of irradiation (1 mw/cm^2), then re-suspended in complete growth medium for measurement of photodynamic inhibition of transport of cycloleucine (7). Alternatively, drug-loaded cell pellets were suspended in 0.5 ml of 50 mM phosphate buffer and extracted twice with 1 ml of ethyl acetate. The organic phases were combined and evaporated for reverse-phase TLC analysis. To measure persistence of intracellular porphyrin, HP-loaded cells were suspended in growth medium for 24 hr, then collected for measurement of remaining intracellular porphyrin by fluorescence and TLC assays.

RESULTS

Table 1 shows levels of HPD and its components which photosensitize L1210 cells *in vitro* so that subsequent irradiation inhibits capacity for active concentration of cycloleucine by 50%. This measurement apparently underestimates the actual kill as shown by viability studies. Purification of HP markedly increased the drug level needed to photosensitize cells. Results of TLC analysis of cells incubated with these drug levels are shown in Table 2.

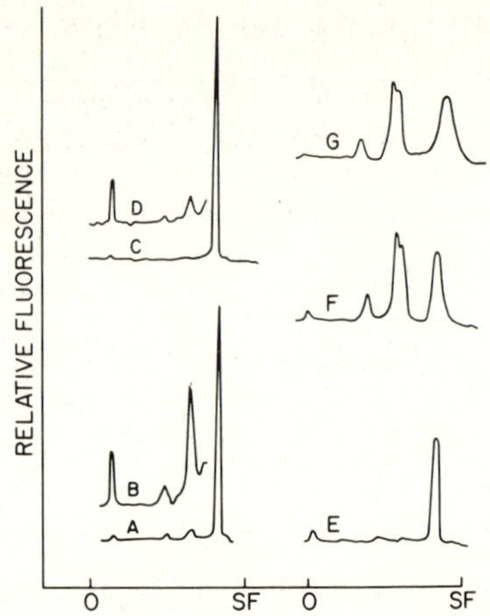

Figure 1. Reverse-phase TLC analysis of (A) commercial HP; (C) purified HP; (B & D) 10X-expanded scans from (A) and (C); (E) cells incubated in 4 μg/ml of purified HP; (F), with 4 μg/ml of unpurified HP; (G), with 60 μg/ml of purified HP. Loading incubations were carried out for 10 min at 37° in serum-free medium.

These results suggest that pure HP is an ineffective photosensitizer of intact cells. Further details are was provided by TLC scans of drugs and cell extracts shown in Fig. 1. The major fluorescent components, best seen on scans F and G, represent (left to right): origin, PP, HVD (2 isomers) and HP. Scan A is an analysis of a solution of HP before purification. Scan C is of purified drug, showing the resulting depletion of HVD and PP. Scans E-G are of extracts of L1210 cells after incubation with 4 μg/ml purified HP (E), 4 μg/ml unpurified HP (F), and 60 μg/ml of purified drug (G). Scans (F) and (G) therefore represent intracellular porphyrin pools formed during incubations of cells with IC_{50} drug levels. The HVD level (scans F and G) was approx that of cells loaded with IC_{50} levels of this porphyrin (Table 2).

Liposomal studies

The suspension of HP-loaded liposomes contained 1 μg of purified HP per ml. Using reverse-phase TLC, we analyzed extracts of cells loaded with HP-containing liposomes after 30 min incubations in serum-free media and after a subsequent 24 hr wash in growth medium following such incubations. The predominant porphyrin, in both cases, comigrated with HP (reverse-phase TLC). The initial drug level achieved was 1 μg/100 mg cells (wet weight). Irradiation (1

Table 2. Photosensitizing ability of porphyrin fractions

Drug fraction	Level μg/ml	Intracellular Level (μg/100 mg cells)		
		HP	HVD	PP
HPD	6	0.09	0.60	0.50
HP(crude)	4	0.12	0.53	0.09
HP(purif.)	60	0.18	0.51	0.08
HVD	1	0.01	0.54	0.11
PP	0.2	<0.01	<0.01	0.52

S-180 cells were incubated with IC_{50} levels of drugs in serum-free growth medium for 10 min at 37°, as specified in the text, and intracellular contents analyzed via reverse-phase TLC. HP(crude) = commercial drug; HP(purif.) was freed from major hydrophobic contaminants by extraction; These results were reproducible to ±20% of values shown in in replicate experiments.

mw/cm², 10 min) of cells loaded with this level of hematoporphyrin via the liposomal preparation resulted in approx. 50% inhibition of capacity for concentrative uptake of radioactive cycloleucine. A subsequent 24 hr wash of the HP-loaded cells resulted in loss of only 20% of total drug. These results are described in more detail in Ref. 14.

Analysis of HPD by reverse-phase TLC (Fig. 2) showed the presence of HP (55%), HVD (25%), PP (5-7%), and other components. We found that none of the three major components of HPD (HP, HVD, PP) could produce long-persisting fluorescence in the mouse Sarcoma 180 solid tumor _in vivo_, while HPD administration (5 mg/kg) resulted in highly fluorescent tumors after 48 hr. TLC studies indicated the major fluorescent compound found in S-180 cells 48 hr after HPD administration migrated with an authentic sample of hematoporphyrin. _In vitro_ photosensitization by HPD, after brief incubations, is presumably mediated by HVD and PP (Table 2). But localization _in vivo_ requires a much longer interval and may be mediated by a different process.

HPD localizing studies

Since tumor localization derives from long-term persistence of porphyrin fluorescence in a cell, retention of HPD

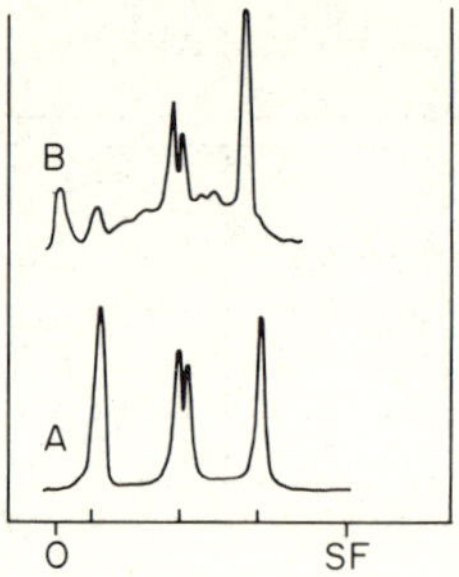

Figure 2. Reverse-phase TLC of (A) standards (left to right: PP, two HVD isomers, HP) and (B) HPD. Fluorescent material can also be seen at the origin of the HPD trace.

and its components by S-180 cells in vitro was examined. For these studies, we incubated S-180 cells for 30 min or 24 hr with different porphyrins, then resuspended the drug-loaded cells in fresh growth medium (containing 10% serum). Results shown in Table 3 may explain the inability of readily-accumulated components of HPD to persist in vivo. Rapid uptake of the hydrophobic components HVD and PP was followed by rapid exodus when drug-loaded cells were suspended in fresh medium. But exposure of cells to HPD resulted in a gradual accumulation of non-diffusible fluorescence. Total porphyrin uptake was not significantly altered when incubations with HPD were prolonged from 0.5 to 24 hr, but the amount of non-diffusible porphyrin fluorescence was markedly increased.

DISCUSSION

HPD is derived from HP via an acetylation reaction followed by alkaline hydrolysis (Fig 3, R=Ac). Additional dehydration reactions, caused by the presence of sulfuric acid in the acetylation mixture, result in formation of HVD and PP (6). These products are also minor contaminants of HP (Fig. 1). There are 2 positional isomers of HVD (one shown) and therefore 2 isomeric monovinyl-monoacetyl derivatives. Two isomeric monoacetyl derivatives of HP are also formed during the acetylation step (16). The alkaline hydrolysis step should result in removal of all acetyl substituents, yielding HPD which, on theoretical grounds, should contain only HP, HVD and PP.

We found that none of the three major components of HPD was a tumor localizer using sub-cutaneously implanted S-180

Table 3. Uptake and Retention Phenomena

Drug	Level μg/ml	30 min loading		24 hour loading	
		no wash	30 min wash	no wash	30 min wash
HPD	30	0.55*	0.18	1.07	0.82
HP	15	0.08	0.02	0.12	0.09
HVD	7.5	0.42	0.08	0.15	0.07
PP	2	0.11	0.04	0.19	0.08

S-180 cells were incubated for 30 min or 24 hr in medium containing 30 μg/ml HPD or the concentration of components present in 30 μg/ml HPD, then washed (if specified), for 30 min in fresh growth medium.

* Total porphyrin level (μg/100 mg cells) was measured by fluorescence assays on cell homogenates in CTAB-buffer.

tumors in the mouse. None of these components was as effective as HPD at creating non-diffusible fluorescence in S-180 cells _in vitro_ (Table 3). These experiments provided an important clue concerning the nature of HPD localization.

In previous reports (7,11), we described the correlation between porphyrin hydrophobicity and photosensitizing capacity in intact cells. HP photosensitization appears to result from preferential uptake of hydrophobic contaminants (HVD and PP) by intact cells (Fig. 1). Hematoporphyrin is clearly a photosensitizer in cell-free systems (12,17,18), but appears to be too hydrophilic to readily cross the cell membrane.

During loading incubations _in vitro_, HVD and PP were readily accumulated by cells, but were readily lost when cells loaded for as long as 24 hr were washed for 30 min in fresh medium at 37° (Table 3). Data shown in Table 3 suggest a slow conversion of one or more HPD components to a non-diffusible fluorescent component which behaves on TLC like HP. We also found that HP loaded into L1210 cells via liposomes was not readily washed out upon subsequent incubation in growth medium for 24 hr. TLC data also suggest that hematoporphyrin is the long-persisting fluorescent product in S-180 cells after treatment of tumor-bearing animals with HPD. It appears that actual uptake of porphyrins is needed

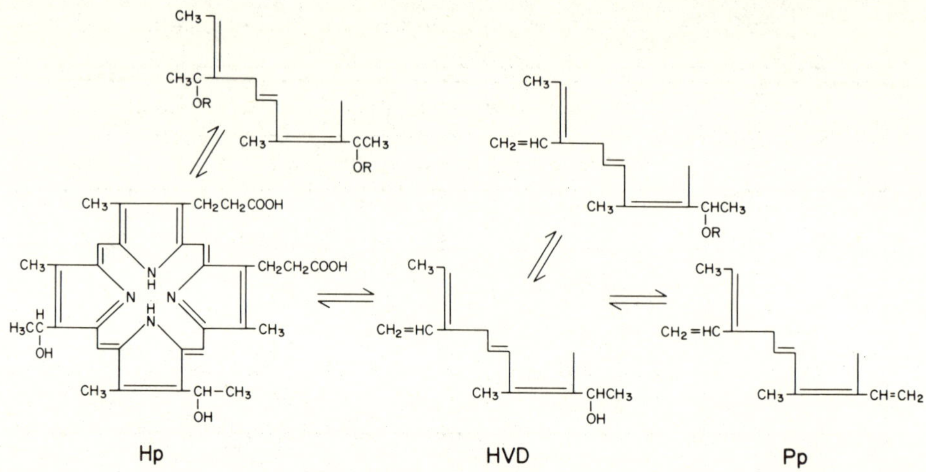

Figure 3. Interactions which occur during the acetylation of HP: both dehydration and acetylation (R=Ac) are shown. The subsequent hydrolysis step converts the acetylated products to <u>sec</u>-alcohols.

for photosensitization of cells (19) or of liposomes (20). This finding may explain the correlation between drug hydrophobicity vs. photosensitizing capacity (7,11).

HP is an effective photo-sensitizer in cell-free systems, but is apparently too hydrophilic to penetrate lipid-rich membrane domains. Photosensitization by HP preparations results from scavenging by cells of hydrophobic contaminants of commercial drug preparations: HVD and PP. An intracellular pool of HP may therefore be unable to diffuse from intact cells, and would represent a long-persisting photosensitizer. Uptake by cells of one or more minor component(s) of HPD, followed by conversion of this material to hematoporphyrin, is therefore proposed to represent the porphyrin-localizing event.

In order to validate this hypothesis, it will be necessary to characterize the structure of the tumor-localizing component(s) of HPD and their intracellular transformations. The proposed hypothesis can account for localization phenomena if conditions necessary for conversion of the postulated HPD component to HP are mainly confined to tumor cells, along with certain host tissues.

CONCLUSIONS

1. Photosensitization of tumor cells is mediated by intracellular porphyrins.

2. The rate of porphyrin uptake is a function of drug hydrophobicity.

3. The aggregation status of intracellular porphyrin is an important determinant of photosensitization.

4. Localization of components of HPD is based on transformation within the cell of a readily permeable porphyrin to a poorly-permeable porphyrin.

5. Fluorescence and absorption studies can be useful for identification of persistently-bound drugs in tumor tissues and loci of such binding.

REFERENCES

1. Cortese, D.A., Kinsey, J.H., Woolner, L.B., Payne, W.S., Sanderson, D.R., and Fontana, R.S. Clinical application of a new endoscopic technique for detection of in situ bronchial carcinoma. Mayo Clin. Proc., 54: 635-642, 1979.

2. Kato, H., Tahara, M., Hayashi, N., Iimura, I., Matsushima, Y., Saito, T., Ono, J., Konaka, C., Aida, M., Hayashi, T., Kawamura, I., and Hayata, Y. Basic studies of diagnosis and treatment of cancer by photoirradiation. Proc. Japanese Cancer Assn., 39: 312 (1980).

3. Dougherty, T.J., Kaufman, J.E., Goldfarb, A., Weishaupt, K.R., Boyle, D., and Mittleman, A. Photoradiation therapy for the treatment of malignant tumors. Cancer Res., 38: 2628-2635, 1978.

4. Dougherty, T.J., Lawrence, G., Kaufman, J.H., Boyle, D., Weishaupt, K.R., and Goldfarb, A. Photoradiation in the treatment of recurrent breast cancer. J. Natl. Cancer Inst. 62: 231-237, 1979.

5. Forbes, I.J., Cowled, P.A., Leong, A.S.Y., Ward, A.D., Black, R.B., Blake, A.J., and Jacka, F.J. Phototherapy of human tumors using haematoporphyrin derivative. Med. J. Aust., 2: 489-492, 1980.

6. Bonnett, R., Ridge, R.J., and Scourides, P. Hematoporphyrin derivative. J. C. S. Comm., 1198-1199, 1980.

7. Kessel, D. Effects of photoactivated porphyrins at the cell surface of leukemia L1210 cells. Biochem., 16: 3443-3449, 1977.

8. Kessel, D., and Kohn, K.I. Transport and binding of mesoporphyrin IX by leukemia L1210 cells. Cancer Res., 40: 303-307, 1980.

9. Kessel, D., and Rossi, E. Determinants of porphyrin-sensitized photooxidation characterized by fluorescence and absorption spectra. Photochem. Photobiol. 35:37-41, 1982.

10. Kessel, D., and Kohn, K.I. Transport and binding of mesoporphyrin IX by leukemia L1210 cells. Cancer Res., 40: 303-307, 1980.

11. Kessel, D. Transport and binding of hematoporphyrin derivative and related porphyrins by murine leukemia L1210 cells. Cancer Res., 41: 1318-1323, 1981.

12. Jori, G., Reddi, E., Rossi, E., Cozzani, I., Tomio, L., Zorat, P.L., Pizzi, G.B., and Calzavara, F. Porphyrin-sensitized photoreactions and their use in cancer therapy. Med. Biol. Envior., 8: 141-154. 1980.

13. Gomer, C.J., and Dougherty, T.J. Determination of $[^3H]$- and $[^{14}C]$ hematoporphyrin derivative distribution in malignant and normal tissue. Cancer Res., 39: 146-151, 1979.

14. Kessel, D. Some observations on components of HPD and their tumor-localizing capacity. Cancer Res., In Press, 1982.

15. Jansons, V.K., Weis, P., Chen, T-H., and Redwood, W.R. In vitro interaction of L1210 cells with phospholipid vesicles. Cancer Res.,38: 531-535, 1978.

16. Clezy, P., Hai, T.T., Henderson, R.W., and Thuc, L.V. The chemistry of pyrrolic compounds XLV. Hematoporphyrin derivative: haematoporphyrin diacetate as the main product of the reaction of haematoporphyrin with a mixture of acetic and sulfuric acids. Aust. J. Chem. 33: 585-597, 1980.

17. Tomio, L., Reddi, E., Jori, G., Zorat, P., Pizzi, G.B., and Calzavara, F. Hematoporphyrin as a sensitizer in tumor phototherapy: effect of medium polarity on the photosensitizing efficiency and role of the administration pathway on the distribution in normal and tumor-bearing rats. Springer Ser. Opt. Sci., 22: 76-82, 1980.

18. Cauzzo, G., Gennari, G., Jori, G., and Spikes, J.D. The effect of chemical structure on the photosensitizing efficiencies of porphyrins. Photochem. Photobiol., 25: 389-395, 1977.

19. Moan, J., and Christensen. T. Cellular uptake and photodynamic effect of hematoporphyrin. Photobiochem. Photobiophys., 2: 291-299 (1981).

20. Suwa, K., Kimura, T., and Schaap, A.P. Reactivity of singlet molecular oxygen with cholesterol in a phospholipid matrix. A model for oxidative damage of membranes. Biochem. Biophys. Res. Commun., 73: 785-791, 1977.

ASPECTS OF THE CELLULAR UPTAKE AND RETENTION OF HEMATOPORPHYRIN DERIVATIVE AND THEIR CORRELATION WITH THE BIOLOGICAL RESPONSE TO PRT IN VITRO

B.W. Henderson, D.A. Bellnier,
B. Ziring and T.J. Dougherty

Roswell Park Memorial Institute
Buffalo, New York

INTRODUCTION

The photosensitizer hematoporphyrin derivative (HPD) is known to be retained selectively in neoplastic tissues, which is the basis for its current use in the diagnosis of early neoplastic lesions, as well as in photoradiation therapy of a wide range of different tumors (1). The reason for porphyrin retention in tumors is not yet known. Porphyrin trapping in the tumor environment due to leaky vascularization and/or poor lymphatic drainage has been suggested as the mechanism (2). However, in view of the selective retention of HPD in very small CIS lesions (3,4), this explanation is not entirely satisfactory. Efforts to detect a special affinity for HPD to malignant cells *in vitro* have yielded controversial results (5-7). Uncertainty also prevails over the cellular target(s) of photoradiation injury. Damage to the cell membrane (8) as well as nuclear damage (9-11) have been reported.

We have carried out some further *in vitro* studies of these questions using two established cell lines (CHO, HeLa) and two primary cell types (EMT-6 tumor cells (12), primary mouse skin fibroblasts (13). HPD or [^{3}H]-HPD were used throughout the experiments.

Fig. 1 shows the uptake of [^{3}H]-HPD by the 4 cell types

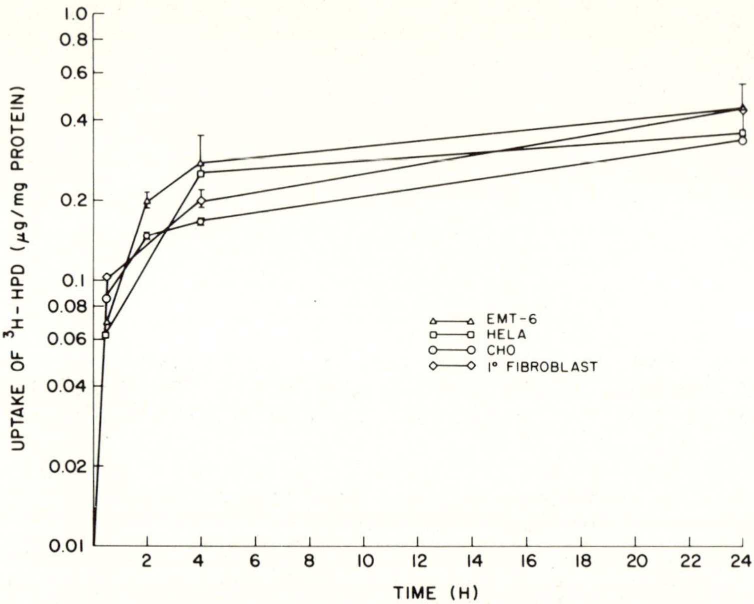

Figure 1. Cells were exposed to 10 µg/ml of [^{3}H]-HPD in growth medium supplemented with 15% NCS, then washed 3X with BSS, scraped off plates, and dissolved in 0.1N NaOH. [^{3}H]-HPD was measured by liquid scintillation counting; protein by the Lowry procedure. Each point is the mean of 2-3 experiments. Three replicate plates were used per point per experiment.

studied. The biphasic pattern of uptake, an initial rapid rise of cellular porphyrin concentration followed by a near-plateau phase, after 3 to 4 hours, as well as cellular porphyrin levels, were almost identical. These data confirm the observation by Chang et al (6) that cells of normal and malignant tissue origin can bind HPD to the same extent _in vitro_.

When the cellular fluorescence of CHO and EMT-6 was compared by cytofluorographic analysis following porphyrin loading for 1 hr and 24 hr respectively, EMT-6 cells showed approximately twice the fluorescence for both time points (Fig. 2). However, when the data were corrected for the different protein content of these cells (CHO: 3×10^{-4} µg protein/cell; EMT-6: 6×10^{-4} µg protein/cell) the HPD bound per cell protein, again, appears to be the same for both cell types. Thus, differences in cell size, i.e. cell protein content, have to be considered when comparisons of porphyrin uptake by different cell types are carried out using individual cell fluorescence only.

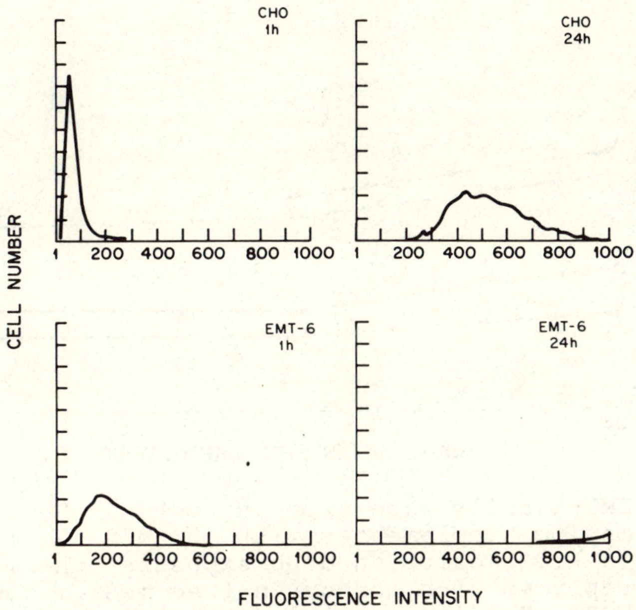

Figure 2. Cytofluorographic analysis of CHO and EMT-6 cells after exposure to 50 μg/ml of HPD for 1 and 24 hr. Cells were exposed to HPD in growth medium containing 15% NCS, detached with 0.25% trypsin, washed twice and suspended in BSS; 10^4 cells were evaluated per sample.

The kinetics of HPD loss from cells upon washing with serum-rich, porphyrin-free medium have been studied in more detail and a representative set of data is shown in Fig. 3 for EMT-6 cells. Cellular porphyrin levels diminish rapidly over a period of 2 to 3 hr, but an apparently 'tightly-bound' HPD portion remains. When corrections are made for cell doubling, this tightly-bound material seems to be retained indefinitely. The portion of tightly-bound HPD increases with increasing time of exposure to the drug over a range from 20% to 70% of initial HPD uptake.

Fig. 4 shows the HPD retention pattern for the 4 different cell types following 30 min and 24 hr exposure times. It is evident that under *in vitro* conditions, all of these cells not only follow the same time course of porphyrin uptake and loss, but also are able to tightly bind a portion of HPD, which according to Kessel (8), corresponds to the hydrophilic components in the HPD mixture.

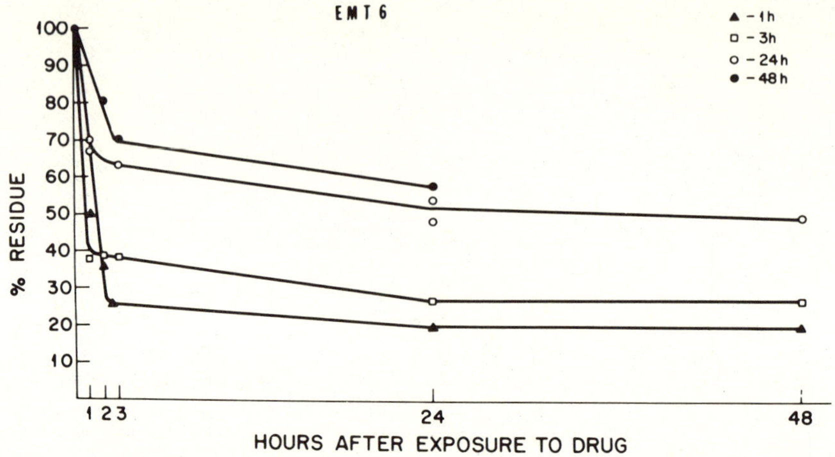

Figure 3. EMT-6 cells were exposed to [^{3}H]-HPD in Waymouth's medium supplemented with 15% NCS for 1, 3, 24 and 48 hr. Following drug exposure, the medium was replaced with porphyrin-free medium containing 15% NCS. Cell-bound [^{3}H]-HPD was assessed as described for Fig. 1. Curves show a single representative experiment with 3 replicate plates per point.

The relative efficacy of 'tightly-bound' and 'loosely-bound' HPD for *in vitro* photoinactivation was examined. Fig. 5 shows the colony-forming ability of CHO cells irradiated with red light following various HPD uptake conditions. Cells incubated in 10 μg HPD/ml for 1.5 hr (loosely-bound HPD) were approximately 3 times less photosensitive than cells containing an identical intracellular concentration of HPD (1.8×10^{-13} μg/cell) but incubated in 25 μg HPD/ml for 12 hr and subsequently washed in serum-rich, porphyrin-free medium for 4 hr (tightly-bound HPD). The same phenomenon was observed for EMT-6 cells where photosensitivity caused by 'tightly bound' HPD was increased by a factor of 3 to 4. These data suggest that long-term uptake conditions result in the migration of HPD or some of its components to sites within the cell whose damage is more critical to cell survival.

The effect of HPD and light on cell lysis was also examined. The release of chromium-51 as an indicator of *in vitro* cell lysis has been described in a number of reports (15,16). Fig. 6 shows that rapid HPD-sensitized photo-lysis varies with both cellular HPD concentration and irradiance in a manner similar to cell lethality (Fig. 7). The relationship between loss of cell viability vs. membrane in-

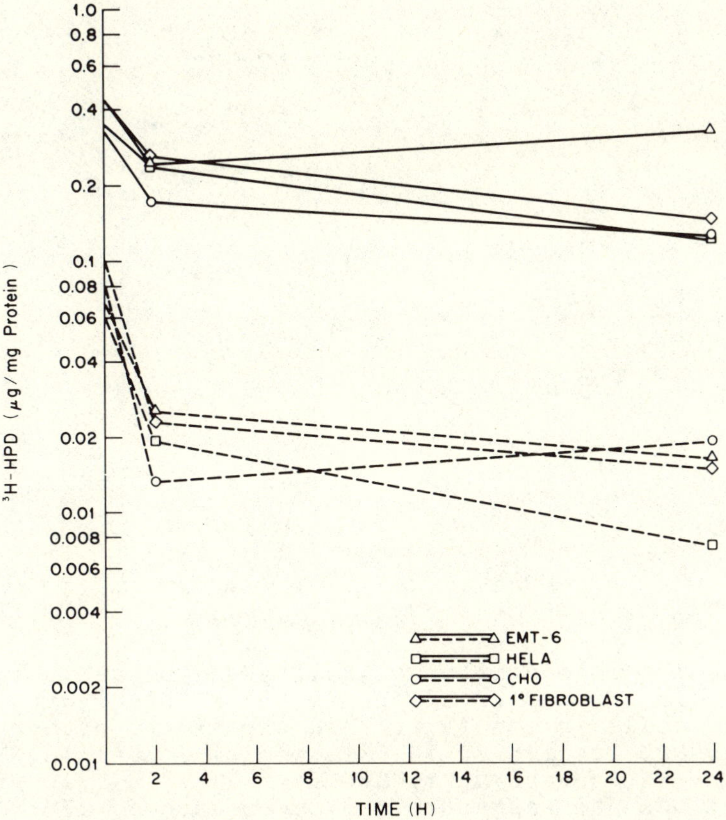

Figure 4. HPD retention by 4 cell types following drug exposure times of 30 min and 24 hr. Cell-bound [^{3}H]-HPD was measured as described for Fig. 1. Curves show one experiment per cell type with 3 replicate plates per point.

tegrity (Fig. 8) shows the onset of lysis (10% ^{51}Cr release) is correlated with onset of exponential cell kill, i.e., the F_q parameter of the survival curve (Fig. 7). Rapid lysis therefore contributes to cell lethality following HPD-sensitized photoirradiation.

REFERENCES

1. T.J. Dougherty, K.R. Weishaupt and D.G. Boyle, Photoradiation Therapy of Malignant Tumors, in: 'Principles and Practice of Oncology', V. DeVita, S. Hellman and S. Rosenberg, eds., J.B. Lippincott Publ. Co., New York (1981).

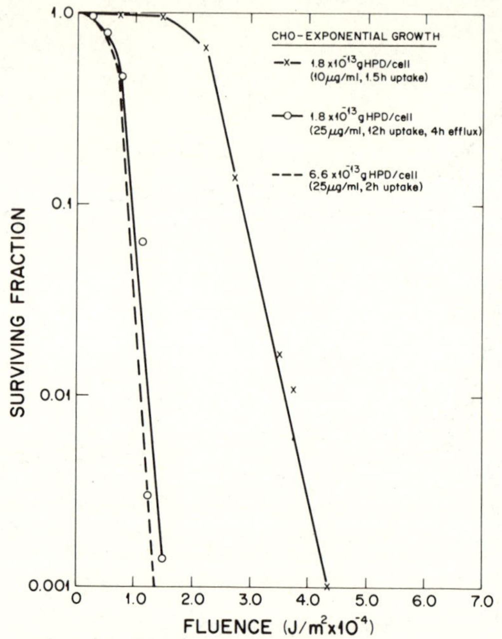

Figure 5. The colony-forming ability of CHO cells incubated in HPD and irradiated with visible light. Single cells in monolayer were exposed to either 10 µg/ml HPD for 1.5 hr in F-10 medium containing 1% NCS (-x-) or 25 µg/ml HPD for 12 hr in F-10 with 1% NCS and washed in medium containing 15% NCS for 4 hr (-o-). Cultures were subsequently illuminated with red light (spectral output: 40W/m², >590 nm) at room temperature, reincubated for 7-10 days and assayed for macrocolony formation. Each point represents the mean of two or more experiments. Three replicate plates were used per point per experiment. Plating efficiencies ranged from 79-90%.

2. P.J. Bugelski, C.W. Porter and T.J. Dougherty, Autoradiographic Distribution of Hematoporphyrin Derivative in Normal and Tumor Tissue of the Mouse, Cancer Res., 41: 4606 (1981).

3. A.E. Profio and D.R. Doiron, Laser Fluorescence Bronchoscope for Localization of Occult Lung Tumors, Med. Phys. 6:523 (1979).

4. D.A. Cortese, J.H. Kinsey, L.B. Woolner, W.S. Payne, D.R. Sanderson and R.S. Fontana, Clinical Application of

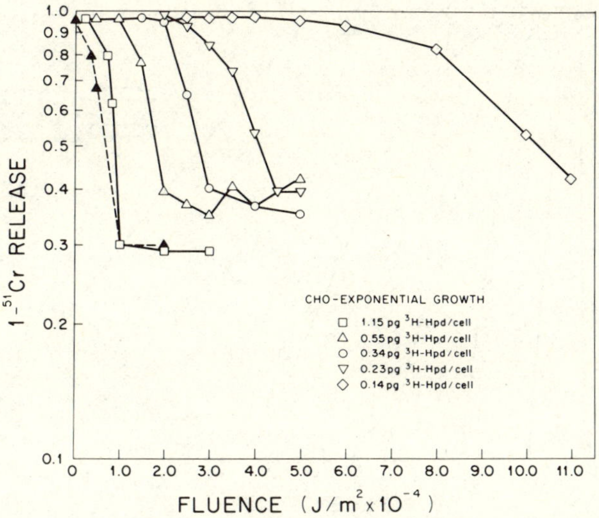

Figure 6. Release of isotopic chromium as a function of fluence (40 W/m^2, >590 nm) for CHO cells containing HPD. Cells were incubated in 50 µCi of ^{51}Cr for 20-22 hr then washed and incubated in F-10 with 1% NBS and 50.00, 25.00, 18.75, 12.50 or 6.25 µg/ml HPD for 2 hr. Intracellular concentrations were 1.15, 0.55, 0.34, 0.23 and 0.14 pg/cell, respectively. The cells were exposed to graded fluences of light and reincubated for 1 hr. In one series, cells incubated with 25 µg/ml HPD (0.55 pg/cell) were irradiated with light and reincubated for 18 hours (▲). ^{51}Cr release was determined using samples of medium containing released Cr. Data are reported in terms of % total cellular ^{51}Cr; each point = mean value of at least 2 replicate experiments; 2-3 dishes/point/determination.

a new Endoscopic Technique for Detection of in situ Bronchial Carcinoma, Mayo Clin. Proc. 54: 635 (1979).

5. B.T. Mossman, M.J. Gray, L. Silberman and R.L. Lipson, Identification of Neoplastic versus Normal Cells in Human Cervical Cell Culture, Obstet. Gynecol. 43:635 (1974).

6. C.T. Chang and T.J. Dougherty, Photoradiation Therapy: Kinetics and Thermodynamics of Porphyrin Uptake and Loss in Normal and Malignant Cells in Culture, Radiat. Res. 74:498 (1978).

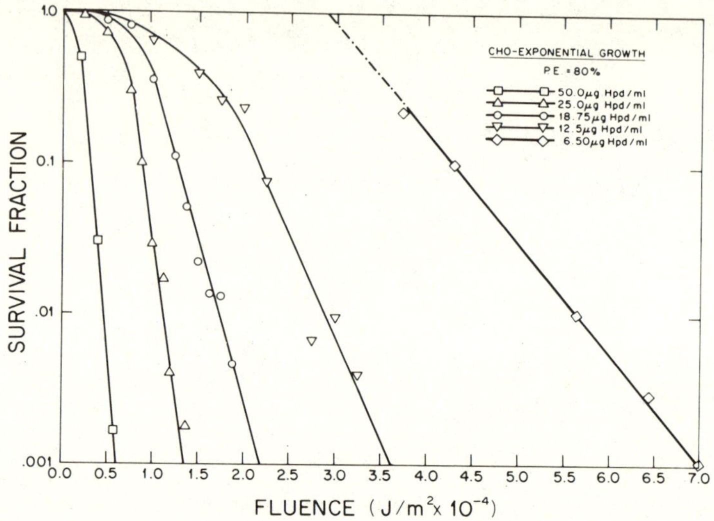

Figure 7. Surviving fraction of CHO cells versus red light fluence (40 W/m², >590 nm). Single cells in exponential growth were incubated(2 hr, 37°, dark) in either 50.00, 25.00, 18.75, 12.50 or 6.50 μg/ml HPD in F-10 medium containing 1% NBS. Irradiated cells were reincubated at 37°C in the dark for 8-10 days, washed, fixed and stained for macrocolony counting. Each point represents the mean of at least 3 experiments with 3 dishes counted per survival point per experiment. Standard deviations (not shown) were typically ±10%.

7. T. Christensen, K. Feren, J. Moan and E. Pettersen, Photodynamic Effects of Hematoporphyrin Derivative on Synchronized and Asynchronous Cells of Different Origin, Br. J. Cancer, in press (1981).

8. D. Kessel, Transport and Binding of Hematoporphyrin Derivative and related Porphyrins by Murine Leukemia L1210 Cells, Cancer Res. 41:1318 (1981).

9. C.J. Gomer, K.R. Weishaupt and T.J. Dougherty, The Effects of Hematoporphyrin Derivative and Visible Red Light on V-79 Cells, Radiat. Res. 74:586 (1978).

10. J. Moan and T. Christensen, Can Porphyrins be used in Diagnosis and Therapy of Cancer? Tumor Res. In Press.

11. D.A. Musser, N. Datta-Gupta and R.J. Fiel, Inhibition of DNA dependent RNA Synthesis by Porphyrin Photosensitizers, Biochem. Biophys. Res. Comm. 97:918 (1980).

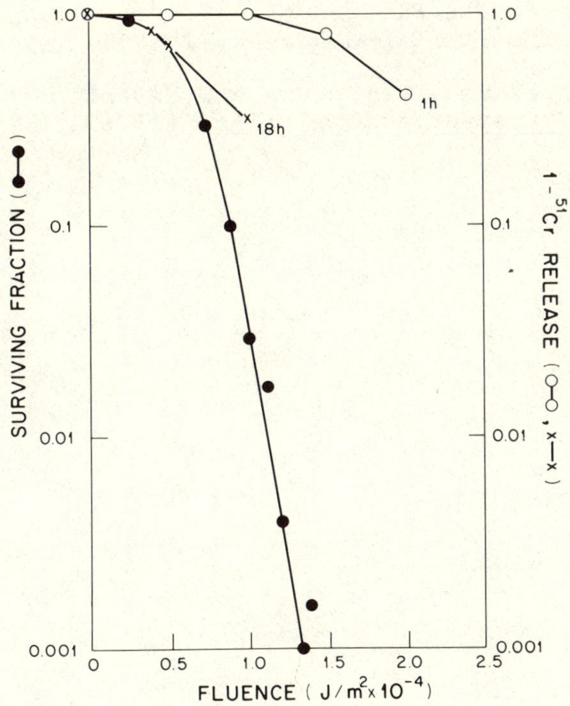

Figure 8. Surviving fraction and isotopic-chromium release as a function of red light fluence (40 W/m², >590 nm) for cells containing 0.55 pg HPD/cell. CHO cells, containing ^{51}Cr and HPD, were exposed to graded fluences of light and reincubated at 37°C for 1 or 18 hr, and ^{51}Cr released was assayed. Cells containing the same concentration of HPD were also exposed to red light, and assayed for reproductive integrity (colony formation) at 8-10 d after irradiation (•).

12. S. Rockwell, In Vivo-in vitro Tumor Systems: New Models for Studying the Response of Tumors to Therapy, Lab. Animal Sci. 27:831 (1977).

13. N.E. Fuseng and P.K.M. Worst, Mouse Epidermal Cell Cultures II. Isolation Characterization and Cultivation of Epidermal Cells from Perinatal Mouse Skin, Expt'l. Cell Res. 93:443 (1975).

14. C.J. Gomer and T.J. Dougherty, Determination of [^{3}H]-and [^{14}C]-Hematoporphyrin Derivative Distribution in Malignant and Normal Tissue, Cancer Res. 39:146 (1979).

15. P.M. Ronai, The Elution of ^{51}Cr from Labelled Leukocytes - A New Theory, Blood 33:408 (1969).

16. C.J. Sanderson, The Uptake and Retention of Chromium by cells, Transplantation 21:256, (1976).

CELLULAR UPTAKE, EXCRETION AND LOCALIZATION OF HEMATOPORPHYRIN DERIVATIVE (HPD)

Michael W. Berns*†, Alan Wile†, Anton Dahlman†, Fred Johnson‡, Robert Burns*†, Daryn Sperling*, Mark Guiltinan*, Ann Siemens*, Robert Walter*, and Marie Hammer-Wilson*

*Department of Developmental and Cell Biology
University of California
Irvine, CA

†Department of Surgery
University of California
Irvine, CA

‡Department of Physics
California State University
Fullerton, CA

INTRODUCTION

Considerable progress has been made towards the development of hematoporphyrin derivative (HPD) plus light as an effective modality in the management of cancer. Recent work in this field largely has been conducted by Dougherty and his colleagues (1-4). Although the method appears to have great potential as a therapeutic modality, many questions are still unanswered. For example, the active ingredients in the HPD solution have not been determined. In addition, the question of whether or not the binding of the HPD and/or its individual subfractions to the outside of the cell and/or to various regions and organelles inside the cell has not been successfully resolved (5-10). The precise mode of cytotoxicity is still unresolved and the mechanism of selective retention by tumor tissue is not understood (10-14).

Our goals have been to approach these questions using *in vitro* as well as *in vivo* models. The present study focuses only on HPD effects on normal cells. It was felt that the effects of HPD on normal cells should be examined before examining HPD effects and retention on malignant cells or the effect of combined HPD + light on malignant cells. This study seemed particularly important since HPD is presumed to have little or no deleterious effects on normal tissue, despite the fact that a patient injected with 3-5 mg/kg body weight of HPD will have a strong hypersensitivity to normal daylight for up to one month following HPD injection.

MATERIALS AND METHODS

HPD. Hematoporphyrin derivative was obtained from Thomas Dougherty of Roswell Park Memorial Institute. Vials containing 5 mg/cc of HPD in saline were stored in the dark until used. All cell treatments reported in this paper used 25 mg/ml of HPD in culture medium containing 10% fetal calf serum.

Cells. Established cell cultures from the rat kangaroo cell line PTK_2 (ATCC#56) were used. These are epithelial cells originally derived from the kidney of a male rat kangaroo *Potorous tridacytlus*. The cells have been maintained in culture for over 10 years during which time they have maintained a near diploid condition of 2n + 1. The cells are contact inhibited, and do not exhibit malignant morphological features.

Cells were grown either in T25 plastic flasks for growth studies (15) or on glass coverslips in multipurpose Rose tissue-culture chambers for fine resolution fluorescent and light microscopy.

Primary cultures of neonatal rat myocardial cells were also used. These cells were established according to standard procedures from 1-3 day old rat ventricles (16). These cells maintain normal morphology and contractile behavior for up to 1 month in culture. For mitochondrial studies, intact myocardial cell mitochondria (3-7 days post establishment) were stained with the fluorescent probe Rhodamine 6G (1.0 mg/ml medium for 30 min. at 37°C proceeded by 3 10 min media rinses). Fluorescence was excited from single Rhodamine 6G-stained mitochondria with a 442 nm He-Cd laser (0.5 u spot diameter).

Growth Studies. Standard growth curves were run on PTK_2 cells grown in T25 flasks. The flasks were each seed with 50,000 cells. Twenty four hr later fresh media containing 25 μg HPD/ml was added and the cells were exposed

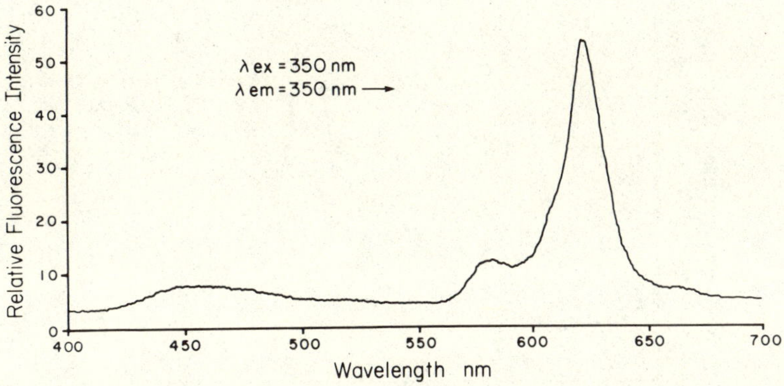

Figure 1. Fluorescence emission spectrum of hematoporphyrin derivative (HPD).

for 18 or 24 hr. Following drug exposure the medium on the cells was changed daily for the next four days and then every other day for the remainder of the experiment. Two flasks were harvested from each group at each medium change and the cells counted on a model ZBi coulter counter. All cell handling was performed in near darkness and the cells were maintained in a 37°C, 5% CO_2 humidified incubator in the dark.

Fluorescence Analysis. Emission and excitation spectra of solutions of HPD in saline were analyzed with a Hitachi-Perkin Elmer fluorescence spectrophotometer MPF3L. Intracellular spectra were obtained using a Nanospec 10 microspectrofluorometer mounted above a Zeiss RA epifluorescence microscope equipped with a 100W mercury lamp. Fluorescence spectra were recorded employing the automatic monochromator drive mounted in front of a photomultiplier tube inside the Nanospec system. The standard Zeiss excitation filter #487709 and band pass filter #BP 570 were employed. These filter combinations permitted excitation at wavelengths below 550nm, and detection of fluorescent emissions above 570 nm. The fluorescence spectra of individual single cells were recorded by positioning them under an aperture in the optical system of the microscope. Emitted fluorescence was projected into the aperture and subsequently to the photomultiplier. Individual fluorescent measurements were also made on PTK_2 cells using a laser-excited fluorescence system (17). This system involves focusing a 442 nm He Cd laser beam to a microspot of 0.5-20 μm in diameter inside a single cell. The fluorescence is detected by a sensitive EMI #9862B photon counting photomultiplier and accumulated by a Tracor Northern #1710 multichannel analyzer. With this system, accurate fluorescence measurements were made on intact cells. This system was also used to record the changes in

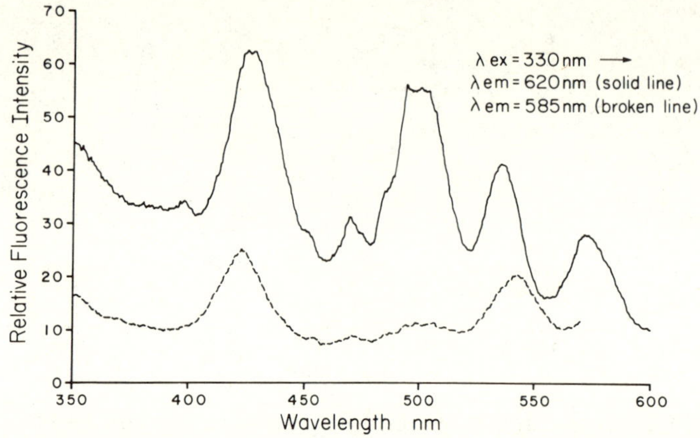

Figure 2. Fluorescence excitation spectra of HPD. Spectra were recorded for the two major emission peaks (see Fig. 1).

fluorescence in the mitochondria of single heart cells that had been treated with both HPD and the respiratory-sensitive fluorochrom Rhodamine 6G (18).

Another system employing a Tracor Northern #1223-3GI diode array was used to study the spatial distribution of HPD fluorescence across an individual cell. This system involved the projection of the fluorescence image of a cross section through the cell onto an array of 1,034 small diodes.

RESULTS

Fluorescence Spectra. HPD solution when examined in cuvettes in the standard fluorometer exhibited a major fluorescent peak at 615-630 nm. A second small peak was observed at 580-590 nm (Fig. 1). When excitation spectra were derived for these two peaks, six distinct excitation peaks were observed for the 615-630 nm emission peak, and only two distinct peaks (425 nm and 540 nm) were detected for the 580-590 nm emission peak (Fig. 2).

The fluorescent emission spectra from within single cells were rather different. A drop of HPD on a slide exhibited three distinct peaks at 610, 640, and 675 nm. Whereas a drop of saline on a piece of filter paper exhibited a distinct red shift of fluorescence with peaks at 620, 645, and 685 nm. The emission spectrum of HPD bound in the cells was blue shifted with three peaks at 600, 635, 670

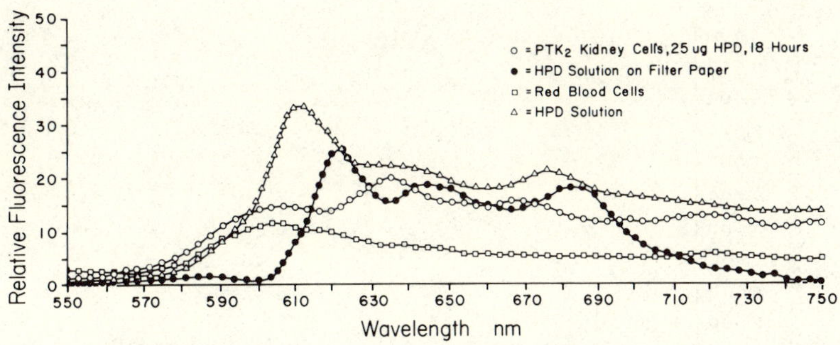

Figure 3. Fluorescence emission spectra of HPD in cells and solution.

nm. In addition the fluorescent peak with the shortest wavelength was always relatively lower in intensity than the two longer wavelength peaks in the cells, whereas in the non-cellular preparation the first peak was always higher in intensity than the two longer wavelength peaks (Fig. 3).

<u>Quantitative Cellular Fluorescence</u>. Fluorescence readings were made on individual cells using a laser-stimulated fluorescence detection system. HPD uptake was determined by measuring fluorescence at 15-20 min. intervals over the first 2 hr following addition of 25 μg/ml of HPD to the culture medium (Fig. 4). There is a rapid uptake over the first 2 hr and a gradual leveling off by 24 hr.

In a subsequent experiment the excretion of HPD was measured starting at 24 hr after the removal of the HPD containing solution. The cells treated with HPD for 18 hr maintained a high level of HPD through 24 hr post-treatment (Fig. 5). By 48 hr post-treatment a substantial loss of HPD from the cells had occurred. The amount of HPD in the cells only decreased slightly over the next 48 hr, and there appeared to be a low level (above control measurements) of HPD maintained within the cells.

<u>Intracellular Localization of HPD</u>. Fluorescent microscopy revealed a distinct orange fluorescence in the cytoplasm with the greatest intensity of fluorescence around the outside of the nucleus. Furthermore, a quantitative reading of fluorescence across the cell with the diode array (Fig. 6) demonstrated distinct perinuclear fluorescence pattern. This pattern correlates with the distribution of lysosomes and mitochondria in these cells.

The effect of HPD on mitochondria was further examined by recording the fluorescence patterns from individual heart

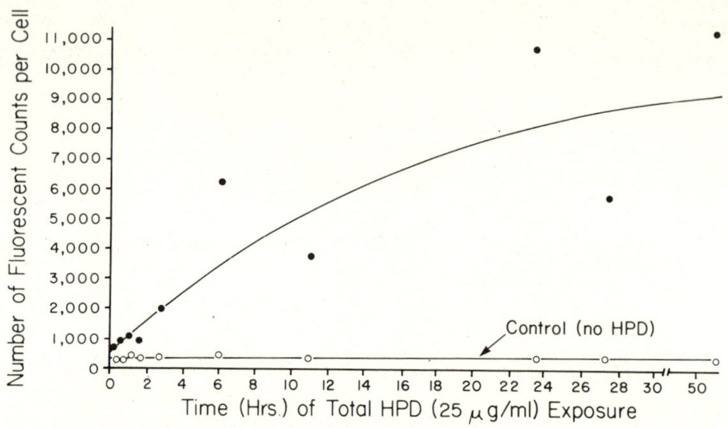

Figure 4. Uptake of HPD by PTK_2 kidney cells as shown by fluorescence measurements on individual cells. Each point represents the mean of at least six different measurements from different cells.

cell mitochondria that had been exposed to the fluorescent probe Rhodamine 6G and HPD. It has been previously shown that treated mitochondria exhibit oscillating fluorescence. Fig. 7a illustrates a normal oscillating fluorescence pattern of Rhodamine 6G stained mitochondrion. Fig. 7b is the fluorescence pattern of a mitochondrion in an HPD-treated cell at 12 minutes following HPD treatment.

Cell Survival. Cells treated with HPD (25 μg/ml) for 18 hr exhibited a definite reduction in growth rate. Most of the cells exposed to HPD (25 μg/ml) for 24 hr were killed by day 9 post-treatment (Fig. 8).

DISCUSSION

Hematoporphyrin derivative in solution has two distinct fluorescence peaks one at 615-635 nm and the other at 580-590 nm. Each of these peaks has a different excitation spectrum suggesting that HPD is a multicomponent solution. When measured on the more sensitive microfluorometer the longer wavelength fluorescence emission could be resolved into three distinct peaks. These peaks demonstrated a blue shift when measurements were taken from inside single cells. Furthermore, the relative proportion of the three peaks was altered when HPD was bound inside the cell. These shifts and changes in intensity may reflect the binding and localization of specific subfractions of HPD.

The quantitative measurements on the cells demonstrated

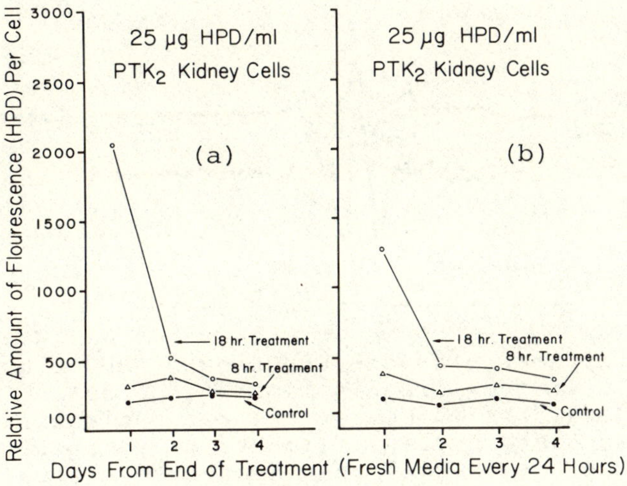

Figure 5. Loss of HPD from PTK_2 Kidney cells. Each point represents the mean of at least six separate measurements in different cells.

a rapid uptake of HPD (or one of its components) by 2 hr of treatment. Significant amounts of a porphyrin component were detected in the cells within 10 minutes of treatment indicating a rapid internalization. The fluorescence microscope also revealed an intracellular pattern of fluorescence surrounding the nucleus. Little or no fluorescence was detected at or near the cell surface or inside the nucleus. This distribution pattern was confirmed by the diode array measurements across the cell.

The studies on the mitochondrial fluorescence patterns demonstrated that HPD rapidly inhibits fluorescence oscillations that are detected by monitoring Rhodamine 6G fluorescence in single mitochondria. These observations further substantiate the suggestion by others (8) that the mitochondria are affected by the internalized fraction of HPD.

Quantitative HPD determinations on cells that have been removed from the HPD containing culture medium demonstrates a loss of HPD from cells by 40 hr after HPD treatment. However, a low level of HPD is retained in cells for up to 4 days post-treatment. Measurements were not made past this time period, but by the slope of the excretion curve, it is likely that a low level of HPD will be maintained in these cells for a considerable period of time.

Growth studies on HPD-treated cells clearly demonstrate that concentrations of HPD at 25 μg/ml for 18 and 24 hr have a growth inhibiting and toxicity effect even though the

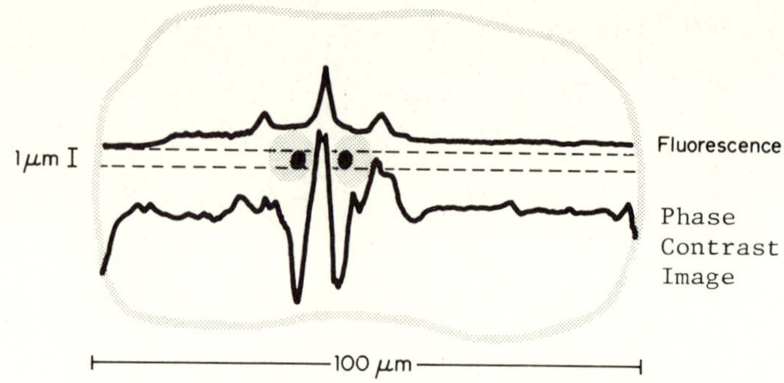

Figure 6. Intracellular distribution of HPD fluorescence. Outer grey line is membrane of cell. Two small dark dots in center are two nucleoli, each in a separate nucleus (lighter grey circular objects). The entire cell was exposed to a fluorescence stimulating mercury lamp. A 1 μm long section (the region between the two dashed lines) through the fluorescing cell was projected onto a diode array. The upper dark line indicates relative fluorescence across the cell. The peaks indicate an increased fluorescence intensity. The lower dark line is an indication of the contrast difference through the same region. A phase contrast image of the same cell region also projected onto the diode array. It was, therefore, possible to match fluorescence with known phase contrast objects.

cells were not exposed to light. These results may not have been detected by other investigators because their cells were simply not followed over a long enough time period, or controls of HPD alone were not run.

The data suggest that HPD definitely affects normal cells without any light addition. It is also evident that even though HPD does come out of normal cells rather rapidly, a small amount is retained. We have to determine if retained HPD photosensitizes cells. Studies on comparison of HPD response by normal vs. malignant cells are underway.

ACKNOWLEDGEMENTS

Supported by AFOSR 80-0062A, NIH P41 RRO 1192, NIH RO1 HL 15740, and NIH RO1 GM 23445.

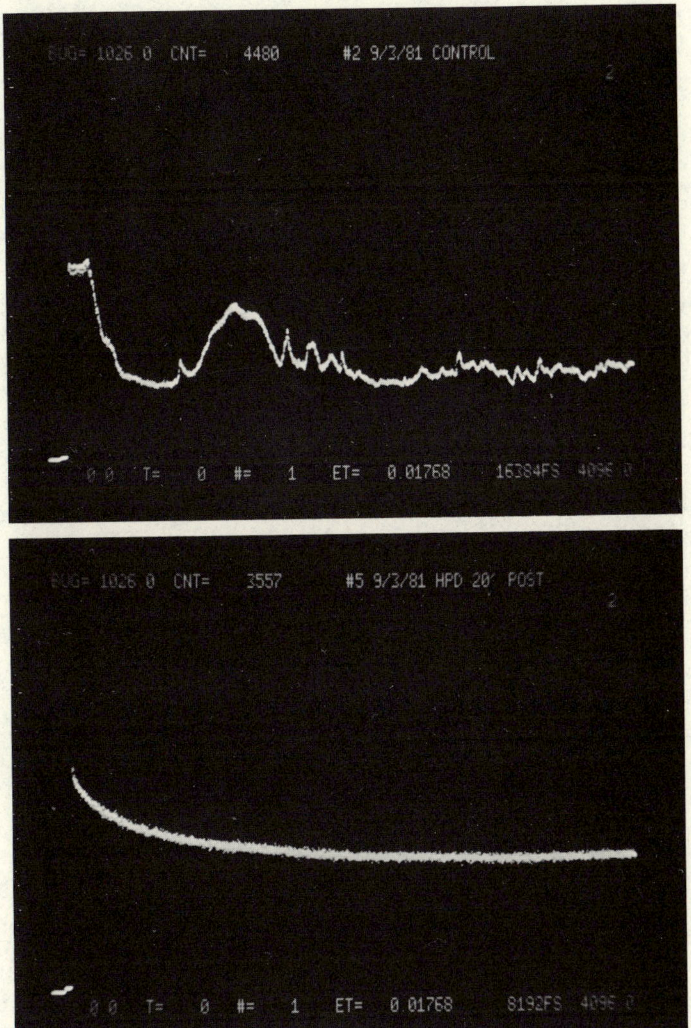

Figure 7 (a). The fluorescence emission at a fixed wavelength of a 1 μm region of a single mitochondrion stained with Rhodamine 6G. Note the oscillating fluorescence pattern. (b) A similar emission taken from a mitochondrion stained with Rhodamine 6G and also treated with HPD, 25 μg/ml; no oscillating patterns observed as in (a).

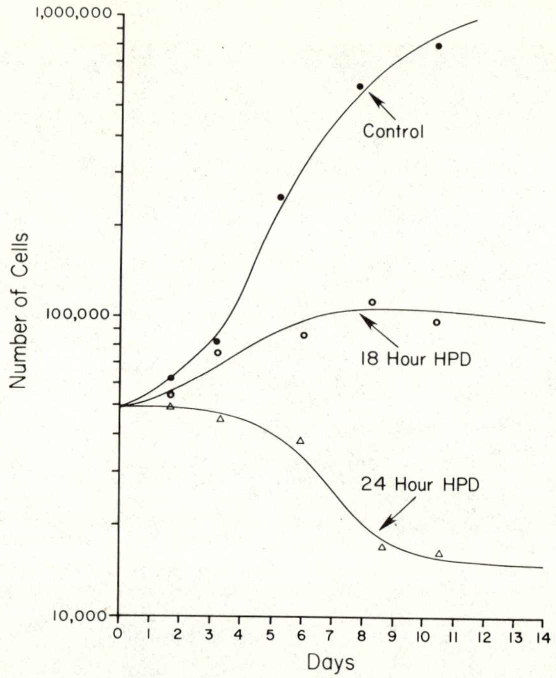

Figure 8. Growth curves of PTK_2 kidney cells following treatment with HPD (25 μg/ml).

REFERENCES

1. Dougherty, T.J. Photoradiation therapy for cutaneous and subcutaneous malignancies. J. Invest. Dermatol. 77:122-124, 1981.

2. Dougherty, R.J., G.B. Grindey, R. Field, K.R.Weishaupt, and D.G. Boyle. Photoradiation therapy. II. Cure of animal tumors with hematoporphyrin and light. J. Natl. Cancer Inst. 55:115-121, 1975.

3. Dougherty, T.J., J.E. Kaufman, F. Goldfarb, K.R. Weishaupt, D. Boyle, and A. Mittleman. Photoradiation therapy for the treatment of malignant tumors. Cancer Res. 38:2628-2635, 1978.

4. Dougherty, T.J., G. Lawrence, J.H. Kaufman, D. Boyle, K.R. and A. Goldfarb. Photoradiation in the treatment of recurrent breast carcinoma. J. Natl. Cancer Inst. 62:231-237, 1979.

5. Christensen, T., and J. Moan. Binding of hematoporphyrin to synchronized cells from the line NHIK 3025. Cancer Lett. 9:105-110, 1980.

6. Coppola, A.. E. Viggiani, L. Salzarulo, and G. Rasile. Ultrastructural changes in lymphoma cells treated with hematoporphyrin and light. Amer. J. Path. 99:175-181, 1980.

7. Kessel, D. Transport and binding of hematoporphyrin derivative and related porphyrins by murine leukemia L1210 cells. Cancer Res. 41:1318-1323, 1981.

8. Salet, C. and G. Moreno. Photodynamic effects of hematoporphyrin on respiration and calcium uptake in isolated mitochondria. Int. J. Radiat. Biol. 39:227-230, 1981.

9. Stenstrom, A.G.K., J. Moan, G. Brunborg, and T. Eklund. Photodynamic inactivation of yeast cells sensitized by hematoporphyrin. Photochem. Photobiol. 32:349-352, 1980.

10. Torinuki, W., T. Miura, and M. Seiji. Lysosome destruction and lipoperoxide formation due to active oxygen generated from hematoporphyrin and UV irradiation. Brit. J. Dermatol. 102:17-27, 1980.

11. Christensen, T., and J. Moan. Photodynamic inactivation of synchronized human cells in vitro in the presence of hematoporphyrin. Cancer Res. 39: 3735-3737, 1979.

12. Dougherty, T.J., C.J. Gomer, and K.R. Weishaupt. Energetics and efficiency of photoinactivation of murine tumor cells containing hematoporphyrin. Cancer Res. 36:2330-2333, 1976.

13. Sery, T. Photodynamic killing of retinoblastoma cells with hematoporphyrin and light. Cancer Res. 39:96-100, 1979.

14. Weishaupt, K.R., C.J. Gomer, and T.J. Dougherty. Identification of singlet oxygen as the cytotoxic agent in photo-inactivation of a murine tumor. Cancer Res. 36:2326-2329, 1976.

15. Siemens, A.E., M.C. Kitzes, and M.W. Berns. Hydrazine effects on vertebrate cells in vitro. Toxicol. Appl. Pharmacol. 55:378-392, 1980.

16. Kitzes, M.C. and M.W. Berns. Electrical activity of rat myocardial cells in culture: La^{+++} induced alterations Amer. J. Physiol.: Cell Physiol. 6:C87-C95, 1979.

17. Berns, M.W. Fluorescent analysis of cells using a laser light source. Cell Biophys. 1:1-13, 1979.

18. Siemens, A.E., R.J. Walter, L.H. Liaw, and M.W. Berns. Laser stimulated fluorescence of submicron regions within single mitochondria of Rhodamine treated myocardial cells in culture. Proc. Natl. Acad. Sci. In Press.

STUDIES OF HPD: CHEMICAL COMPOSITION AND IN VITRO PHOTOSENSITIZATION

Terje Christensen, Johan Moan, James Brown McGhie, Helga Waksvik and Hein Stigum

Norsk Hydro's Institute for Cancer Research
Montebello
Oslo 3, Norway

INTRODUCTION

Photoactivated porphyrins are at present being evaluated for use in cancer therapy (Kelly and Snell, 1976; Dougherty et al., 1978). The method is based on two properties of particular porphyrins: Their tumor-localizing ability (Gomer and Dougherty, 1979) and their photodynamic effect (for a review see Moan and Christensen, 1980). Porphyrin photosensitization is also manifested in the skin of patients suffering from porphyrias.

The mechanisms leading to photosensitization are not well understood. In clinical use of photoactivated hematoporphyrin derivative (HPD) for tumor destruction, a rapid necrotization of treated tumor areas is often observed (Kelly et al., 1975). It is not known at present whether this rapid effect is caused by direct attack of the main cytotoxic product, singlet oxygen (Moan et al., 1979), on the tumor cells or by other factors. A recent report described physiological changes in transplantable C3H mouse mammary tumors after treatment with HPD and light (Bicher et al., 1981). The conclusion was that the treatment induced hypoxia in the tumor tissue without any dramatic change in pH. In view of the results of Dougherty et al., (1981) who found an accumulation of HPD in the vascular stroma, these data suggest damage to the microcirculation which may lead to hypoxia.

On the cellular level, a variety of effects of photosensitizing compounds have been reported (for a review see Amagasa, 1981). Interaction between the photosensitizer and the cell may lead to different cellular effects for photosensitizers with different binding sites (Ito, 1980). In the case of porphyrins, binding sites with different characteristics have been identified (Kessel and Kohn, 1980). Binding to different sites seems to correlate with the hydrophobicity of the porphyrin as well as with the time of exposure of cells to porphyrins (Kessel, 1981; Sandberg and Romslo, 1981). In this study, we will show that HPD contains components with widely different abilities to sensitize cells to light.

Several effects of photoactivated porphyrins upon the cell membrane and cell surface have been observed. These include crosslinking of membrane proteins, inhibition of transport of essential components, perturbation of permeability barriers, inactivation of membrane enzymes, leakage of enzymes out of the cell and increased uptake of porphyrins (Christensen et al., 1981; Kessel, 1977; Kohn and Kessel, 1979; Moan and Christensen, 1981). Subsequent to membrane damage, swelling and lysis of cells have been observed (Moan et al., 1979). Damage to other constituents of cells are also observed after porphyrin and light treatment. Of particular interest are the observations of DNA damage (Gomer, 1980; Moan et al., 1980) and impaired protein synthesis (Wakulchik et al., 1980).

It is therefore likely that several cell biological parameters can be altered by photoactivated porphyrins. It was recently shown that NHIK 3025 cells were killed by two different mechanisms (Christensen, 1981): a rapid lysis of the cells after light exposure, and a partly-reversible inhibition of cells in mitosis. This paper will show the dose-effect relationships of these two modes of photodynamic inactivation.

MATERIALS AND METHODS

Cell Cultivation

Cells from the NHIK 3025 line were cultured in monolayer as described previously (Nordbye and Oftebro, 1969; Oftebro and Nordbye, 1969; Pettersen et al., 1973). The growth medium, E2a, contained 20% human and 10% horse serum. For assay of sister chromatid exchanges (SCE) cells routinely cultivated in MEM (Gibco) with 10% newborn calf serum were used.

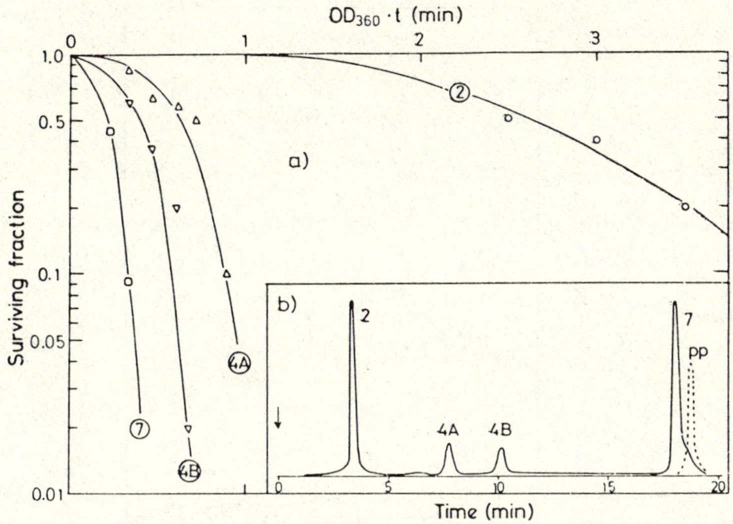

Figure 1. a) Survival of NHIK 3025 cells after treatment with the components of HPD and light. Cells were incubated at 37°C for 1 hr in contact with HPD components dissolved in PBS.
b) Separation of HPD components by reverse-phase HPLC (Moan and Sommer, 1981). Elution profile of protoporphyrin is shown by the dashed line.

Porphyrins

Hematoporphyrin derivative was prepared from hematoporphyrin dihydrochloride according to the method of Lipson et al. (1961) and stirred for 1 hr at room temperature in 0.1 N NaOH before neutralization and sterilization. The cells were kept in contact with HPD 30 min before illumination.

HPD was analysed and components were isolated by isocratic high pressure liquid chromatography (Moan and Sommer, 1981). The chromatograph (LDC) was equipped with a variable wavelength detector (LDC spectromonitor III, model 1204) and used a Supelcosil LC-18 column. The mobile phase was methanol/water (82:18) supplemented with 1 mM phosphate (pH 5.7 or 8.5). After collection of components, the mobile phase was evaporated under reduced pressure and the porphyrins dissolved in PBS before addition to cells.

Illumination

Cells were subjected to 11.0 W/m² of 350-400 nm light while adhering to Falcon tissue culture flasks. After HPD

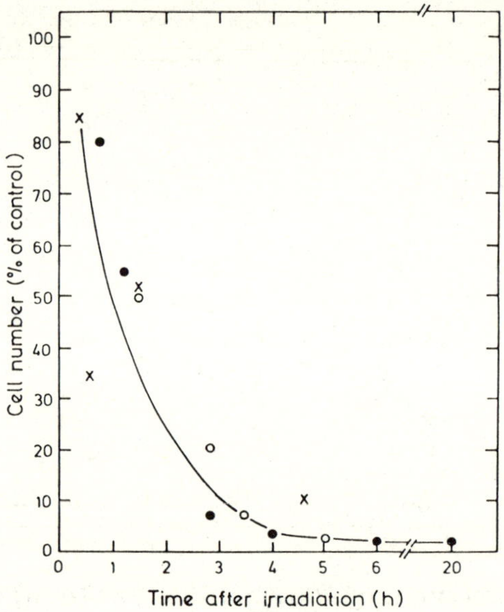

Figure 2. Cell counts after treatment with HPD (0.25 mg/ml) and light. The cells were counted in a volume spectrometer with Coulter flow configuration. After irradiation, the cells were incubated for the indicated time in growth medium at 37°C in the dark. Different symbols represent independent individual experiments.

and light treatment, the cells were incubated 1 week before staining and counting of colonies.

Cell lysis was assayed by two different methods. The first method involved counting intact cells with a phase contrast microscope (Leitz Diavert); 2-5 x 10^4 cells were placed in Falcon tissue culture flasks and allowed to attach. The number of cells not lysed by HPD + light was compared to controls. It was assumed that no cells detached from the substratum 24 hr after treatment.

The second method was based on quantification of the cell number in a laboratory-built volume spectrometer with Coulter flow configuration (Steen and Lindmo, 1978). Approx. 3 x 10^5 cells were inoculated and treated as above. The cells were removed from the flasks with a trypsin-EDTA solution and counted in the volume spectrometer with set volume cut-off values (channels 60-230, Fig. 3).

Sister Chromatid Exchanges

The assay for SCE's has been described previously (Moan

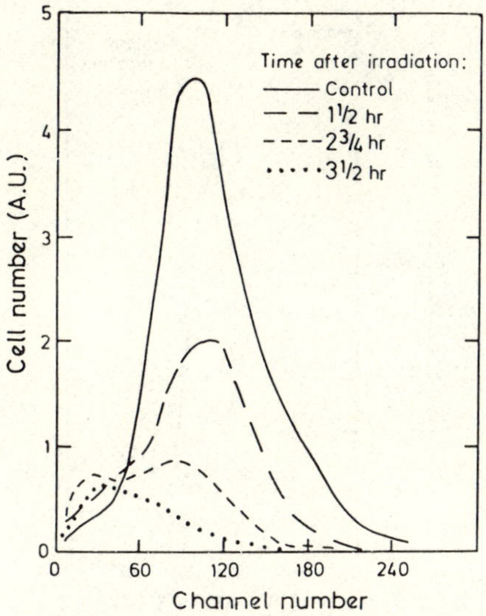

Figure 3. Volume spectra from one of the experiments described in the legend to Fig. 2. Cellular volume is proportional to channel number.

et al., 1980). In this part of the study, cells routinely cultivated in MEM-Eagle's medium were used. The cells were treated with 25 μg/ml HPD in MEM, with 0% or 10% serum, for 30 min at 37°C; gas phase 5% CO_2. After light exposure, BrdU was added, and the cells incubated for 50 hr at 37°C prior to staining. Unlike a previous study (Moan et al., 1980) the present study involved treatment with HPD and light _prior to_ incorporation of BrdU. This was necessary because photoactivation of BrdU induced an increased frequency of SCE at the highest light doses applied here.

RESULTS

In Fig. 1a, dose-response curves for cells sensitized in PBS with each of the 4 main compounds of HPD (Fig. 1b) are shown. The lysis of cells treated with HPD and a relatively large light dose proceeds as shown in Figs. 2 and 3. During 4 hr after treatment, the cell number (recorded by Coulter counting) is gradually reduced. Figs. 4 and 5 show dose response curves for cell inactivation (loss of colony forming ability) and cell lysis. In Fig. 4, the cells were treated under low serum conditions and in Fig. 5, 30% serum was present. These data show that the dose response curve for lysis has a more pronounced shoulder than that for cell inactivation. Fig. 6 was constructed from

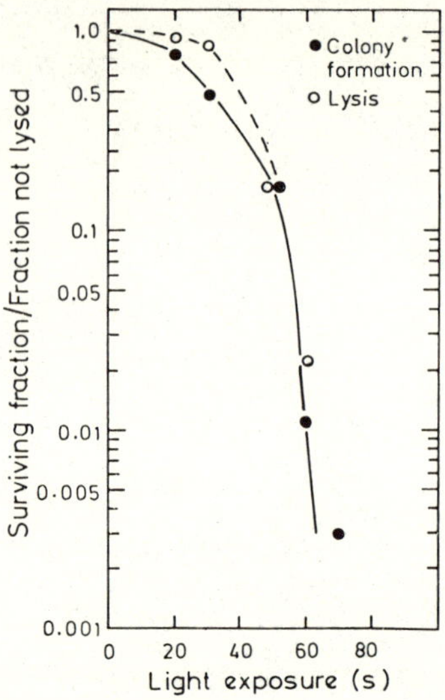

Figure 4. Inactivation (solid line) and lysis (dashed line) of cells treated with 25 μg/ml of HPD in phosphate-buffered saline containing 1% human serum, then exposed to light. Inactivation was measured by loss of the ability to form macroscopic colonies in vitro. Lysis was assayed by microscopic observation of the cells 24 hr post-treatment. Each point is the mean value of 2-4 individual experiments each performed with 2-3 replicate flasks.

data like those shown in Figs. 4 and 5, and shows cell inactivation as a function of cell lysis. With both high and low serum concentration during treatment, more than 60% of the cells were inactivated before lysis was observed. Furthermore, Fig. 6 indicates that cells treated under low serum conditions were more prone to lysis than cells treated in the presence of 30% serum. The frequency of SCE's was also dependent on the serum concentration during treatment with HPD and light. When compared at the same level of survival, fewer SCE's are induced when cells are treated in the absence of serum than in the presence of 10% serum (Fig. 7).

DISCUSSION

Under the conditions used in this study, the most hydrophobic component caused the highest degree of photoinactivation of NHIK 3025 cells (Fig. 1). In another study

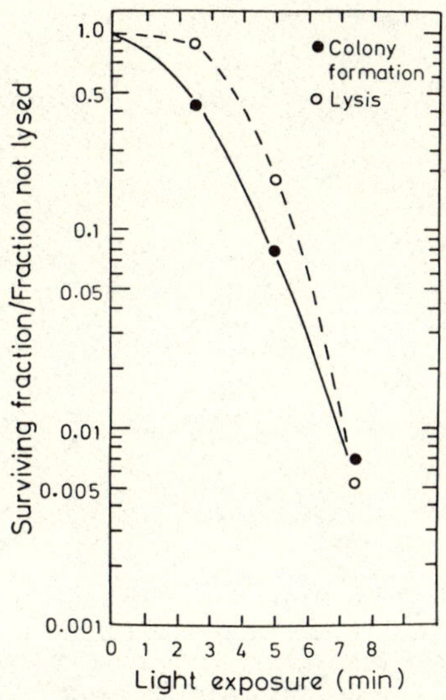

Figure 5. Inactivation and lysis of cells treated with 0.25 mg/ml of HPD in growth medium containing 30% serum. Symbols and method as described in the legend of Fig. 4.

performed in this laboratory, it was concluded that the main photosensitizing component of HPD, as well as of crude hematoporphyrin (HP) treated with alkali, was component 7 (Moan et al., in press). These data correspond well with the findings of Kessel (1981). Kessel observed that the most hydrophobic components were readily washed from cells by medium with serum, even after several hours of contact between cells and porphyrin. This was not the case for the hydrophilic fractions which formed stable bonds with some cellular structures. Therefore, the hydrophilic fractions may also play an important role in the photosensitization of tumors. These questions are presently being investigated in our laboratory.

Protoporphyrin had a slightly longer retention time on the HPLC column than component 7 (Fig. 1b). We therefore assume that the major portion of component 7 is not protoporphyrin. The nature of these components is more thoroughly discussed elsewhere (Moan et al., 1981).

The loss of plasma membrane integrity and necrosis is an early and dramatic event after exposure of cells and tissues to porphyrins and light (Moan et al., 1979; Kelly et al., 1975). But cells may tolerate some damage without inactivation (Dougherty et al., 1981) and they may be inac-

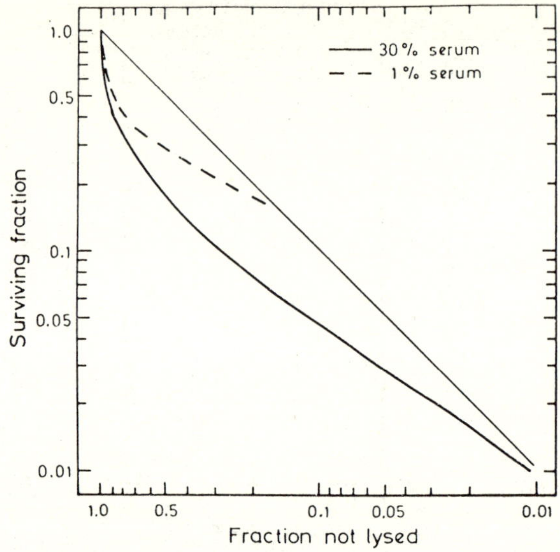

Figure 6. Relation between surviving fraction and the fraction of cells not lysed by 24 hours post treatment. The curves have been constructed from the data shown in Figs. 4 and 5. Solid line represents cells treated in 30% serum and dashed line cells treated in 1% serum. The narrow line indicates a hypothetical curve for a treatment modality inducing cell lysis and cell inactivation at equal rates.

tivated without immediate lysis by treatment with photoactivated HPD (Christensen, 1981). Inactivation without initial lysis occurs when treated cells are arrested in mitosis for several hours. This mitotic inhibition is probably due to a block in the cell cycle between metaphase and anaphase (Christensen, 1981). Figs. 4, 5 and 6 show that cell lysis becomes important only at high light doses. At lower doses, which inactivate up to about 60% of the cell populations, no cells lose their membrane integrity for 24 hr after treatment. When the dose was increased, the membrane damage became severe, and all inactivated cells were readily lysed. Kinetics of the lysis mechanism can be seen in Figs. 2 and 3. Most cells were lysed 4 hours after treatment. This was accompanied by the appearance of small volume particles, probably representing debris resulting from lysis. In a previous report, we found that enzymes leaked out of the cells at approximately the same rate as cell lysis proceeded (Christensen et al., 1981).

A recent report, describing a dual phototoxic effect of

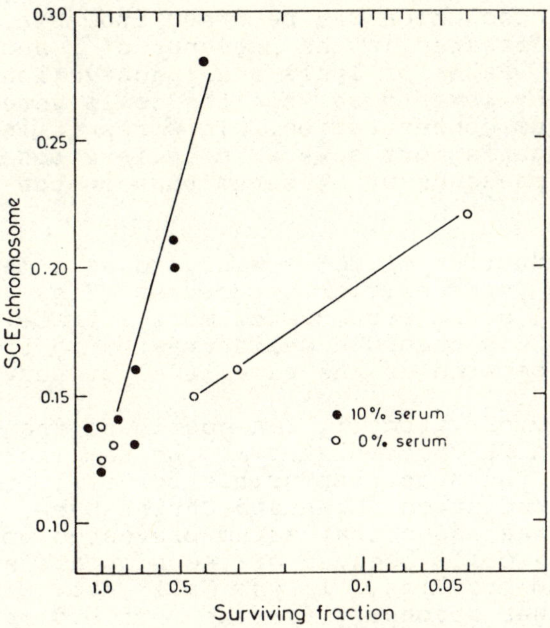

Figure 7. Frequency of SCE's as a function of decreasing survival for cells treated with 25 μg/ml of HPD in MEM either with 10% (solid symbols) or without (open symbols) serum. Each symbol is the frequency determined in a separate experiment performed with two replicate flasks. Survival was determined by treating the same cell populations in 2-3 independent experiments and assaying colony forming ability of the cells.

protoporphyrin, supports the present findings (Ortner et al., 1981). Rat peritoneal mast cells were found to lose the ability to react to a stimulator of histamine secretion after low doses of photoactivated protoporphyrin. At higher doses, histamine was readily released by cell lysis. A similar phenomenon was observed by testing the osmotic fragility of erythrocytes. At low doses a decreased fragility was observed followed by a progressive increase in osmotic fragility at higher doses.

Studies involving the measurement of photodynamic effect by cell lysis (e.g. trypan blue permeability) or necrosis of tumor tissue should take the above observations into account. The presence of apparently normal tumor tissue after treatment (Kelly et al., 1975) will not necessarily imply regrowth of the tumor.

Another conclusion may be drawn from Fig. 6 by comparing the cells treated in the presence of 1 and 30% serum,

Another conclusion may be drawn from Fig. 6 by comparing the cells treated in the presence of 1 and 30% serum, respectively. Rates of lysis and inactivation became equal at a relatively lower dose when the cells were treated with the lower serum concentration. This result indicates that membrane damage is more severe in cells exposed to HPD and light in the presence of 1% serum than in the presence of 30% serum.

The induction of SCE's was used as a means of assaying damage to intracellular structures (Fig. 7). Cells treated in 10% serum accumulated more intracellular damage in terms of SCE's than did cells treated in the absence of serum, when compared at the same level of survival.

In a previous study it was concluded that 10^{-4} M hematoporphyrin in PBS accumulated at the same rate in the nucleus as at the site responsible for cell inactivation after photoirradiation (Moan and Christensen, 1981). Furthermore, it was shown that serum prevented uptake of hematoporphyrin by cells because of the high affinity of porphyrins for serum proteins. It was later shown that the binding to serum was strongest for the most hydrophobic component of HPD (Moan and Sommer, 1981). Hydrophobic porphyrins have been shown to cause damage mainly in membrane structures upon photoactivation (Sandberg and Romslo, 1981). One may speculate that the amount of photoinduced membrane damage observed in cells labelled with HPD in the presence of high concentrations of serum is reduced because the uptake of component 7 is preferentially inhibited by serum proteins.

Our data do not necessarily indicate the absence of membrane damage at low doses of porphyrins and light. It has been shown that damage to cells by this mode of therapy is repairable (Christensen, unpublished results; Weishaupt et al., 1979; Schothorst et al., 1980). This means that cells may accumulate some membrane damage without lysis. Cell death without prior lysis may be a secondary effect of the membrane damage. The variety of transport processes that are damaged by porphyrins and light may lead to substantial disturbances of cell metabolism (Kessel, 1977). Photooxidation of other targets than the cell membrane should also be evaluated as primary events in photodynamic inactivation.

Mitochondrial and lysosomal enzyme systems are affected by photoactivated porphyrins (Sandberg and Romslo, 1981). Furthermore, RNA-polymerase is highly sensitive to photodynamic damage (Munson, 1979). Single strand breaks in DNA, mutation (Gomer, 1980), sister chromatid exchanges and chromosome aberrations (Evensen and Moan, in press) were observed after HPD and light treatment of cells *in vitro*. The role of these effects on cell kill has not yet been determined.

Data presented here suggest that HPD and light-induced damage leads to different modes of cell death. The mechanism of rapid lysis which follows photoradiation is most probably related to disturbances of membrane permeability. Cell inactivation mechanisms by mitotic inhibition and other subcellular mechanisms should be studied more thoroughly to determine whether these mechanisms play a role in eradication of tumors *in vivo*.

CONCLUSIONS

Human NHIK 3025 cells were exposed to HPD + light. Cellular responses were observed and the following conclusions were drawn.

1. The most hydrophobic component of HPD had the strongest photosensitizing effect when cells were exposed to light in the presence of photosensitizer. This hydrophobic component is believed to consist of porphyrin aggregates in aqueous solution.
2. Cells incubated with HPD and irradiated with high light doses were lysed within 4 hr.
3. At lower light levels, lysis was not the immediate cellular response leading to inactivation.
4. Membrane damage was more prominent when a low concentration of serum was present during the incubation with HPD.
5. Intracellular damage, estimated from the frequency of sister chromatid exchanges, was more severe in the presence of high serum levels.

ACKNOWLEDGEMENTS

This study was supported by the Norwegian Cancer Society (Landsforeningen mot Kreft) and the Norwegian Research Council for Science and the Humanities.

REFERENCES

1. Amagasa, J. (1981) Dye binding and photodynamic action. *Photochem. Photobiol.* *33*, 947-955.

2. Bicher, H.I., Hetzel F.W., Vaupel P. and Sandhu T.S. (1981) Microcirculation modification by localized microwave hyperthermia and hematoporphyrin phototherapy. *Biblthca. Anat., No 20* (Proc. 11th Europ. Conf. Microcirculation, P. Gaehtgens, ed, 1980), pp. 628-632. Kager, Basel.

3. Christensen, T. (1981) Multiplication of human NHIK 3025 cells exposed to porphyrins in combination with light. Br. J. Cancer 44, 433-439.

4. Christensen, T., Volden G., Moan J. and Sandquist T. (1981) Release of lysosomal enzymes and lactate dehydrogenase due to hematoporphyrin derivative and light irradiation of NHIK 3025 cells in vitro. Ann. Clin. Res., in press.

5. Dougherty, T.J., Kaufman, J.E., Goldfarb, A., Weishaupt, K.R., Boyle, D. and Mittleman, A. (1978) Photoradiation therapy for the treatment of malignant tumors. Cancer Res. 38, 2628-2635.

6. Dougherty, T.J., Henderson, B.W., Bellnier, D.A., Weishaupt, Michalakes, C., Ziring, B. and Chang, C. (1981) Preliminary information pertaining to mechanisms in hematoporphyrin derivative phototherapy of malignant tissue. Abstr, 9th meeting, Am. Soc. Photobiol, p. 124.

7. Evensen, J.F., and Moan, J. (1982) Photodynamic action and chromosomal damage; a comparative study of haematoporphyrin derivative (HPD) with light and X-irradiation. Br. J. Cancer in press.

8. Gomer, C.J. (1980) DNA damage and repair in CHO cells following hematoporphyrin photoradiation. Cancer Lett. 11, 161-167.

9. Gomer, C.J. and Dougherty T.J. (1979) Determination of ^{3}H- and ^{14}C-hematoporphyrin derivative distribution in malignant and normal tissue. Cancer Res. 39, 146-151.

10. Ito, T. (1980) The dependence of photosensitizing efficacy of acridine orange and toluidine blue on the degree of sensitizer cell interaction. Photochem. Photobiol. 31, 565-570.

11. Kelly, J.F. and Snell, M.E. (1976) Hematoporphyrin derivative: A possible aid in the diagnosis and therapy of carcinoma of the bladder. J. Urol. 115, 150-151.

12. Kelly, J.F., Snell, M.E. and Berenbaum, M.C. (1975) Photodynamic destruction of human bladder carcinoma. Br. J. Cancer 31, 237-244.

13. Kessel, D. (1977) Effects of photoactivated porphyrins at the cell surface of leukemia L1210 cells. Biochem. 16, 3443-3449.

14. Kessel, D. (1981) Transport and binding of hematoporphyrin derivative and related porphyrins by murine leukemia

L1210 cells. Cancer Res. 41, 1318-1323.

15. Kessel, D. and Kohn, K.I. (1980) Transport and binding of mesoporphyrin IX by leukemia L1210 cells. Cancer Res. 40, 303-307.

16. Kohn, K. and Kessel, D. (1979) On the mode of cytotoxic action of photoactivated porphyrins. Biochem. Pharmacol. 28, 2465-2470.

17. Lipson, R., Baldes, E. and Olsen, A. (1961) The use of a derivative of hematoporphyrin in tumor detection. J. Natl. Cancer Inst. 26, 1-8.

18. Moan, J. and Christensen, T. (1980) Porphyrins as tumor localizing agents and their possible use in photochemotherapy of cancer. Tumor Res. 15, 1-10.

19. Moan, J. and Christensen, T. (1981) Photodynamic effects on human cells exposed to light in the presence of hematoporphyrin. Localization of the active dye. Cancer Lett. 11, 209-214.

20. Moan, J., Christensen, T., and Sommer, S. (1982) The main photosensitizing components of hematoporphyrin derivative. Cancer Lett., in press.

21. Moan, J., Pettersen, E.O. and Christensen, T. (1979) The mechanism of photodynamic inactivation of human cells in vitro in the presence of haematoporphyrin. Br. J. Cancer 39, 398-407.

22. Moan, J., Sandberg, S., Christensen, T. and Elander, S. (1981) Hematoporphyrin derivatives, chemical composition, photochemical and photosensitizing properties. (This book).

23. Moan, J. and Sommer, S. (1981) Fluorescence and absorption properties of the components of hematoporphyrin derivative. Photobiochem. Photobiophys. 3, 93-103.

24. Moan, J., Waksvik, H. and Christensen, T. (1980) DNA single strand breaks an sister chromatid exchanges induced by treatment with hematoporphyrin and light or by x-rays in human cells of the line NHIK 3025. Cancer Res., 40, 2915-2918.

25. Munson, B.R. (1979) Photodynamic inactivation of mammalian DNA-dependent RNA polymerase by hematoporphyrin and visible light. Int. J. Biochem. 10, 957-960.

26. Nordbye, K., and Oftebro, R. (1969) Establishment of four new cell strains from human uterine cervix I. Exp.

Res. 58, 459.

28. Ortner, M.J., Abhold R.H. and Chignell, C.F. (1981) The effect of protoporphyrin on histamine secretion by rat peritoneal mast cells: A dual phototoxic reaction. Photochem. Photobiol. 33, 3355-360.

29. Pettersen, E.O., Oftebro, R. and Brustad, T. (1973) X-ray inactivation of human cells in tissue culture under aerobic and extremely hypoxic conditions in the presence and absence of TMPN. Int. J. Radiat. Biol. 24, 285-296.

30. Sandberg, S. and Romslo, I. (1981) Porphyrin-induced photodamage at the cellular and the subcellular level as related to the solubility of the porphyrin. Clin. Chim. Acta 109, 193-201.

31. Schothorst, A.A., De Haas, C.A.C. and Suurmond, D. (1980) Photochemical damage to skin fibroblasts caused by protoporphyrin and violet light. Arch. Dermatol. Res. 268, 31-42.

32. Steen, H.B. and Lindmo, T. (1978) Cellular and nuclear volume during the cell cycle of NHIK 3025 cells. Cell Tissue Kinet. 11, 69-81.

33. Wakulchik, S.D., Schiltz, J.R. and Bickers, D.R. (1980) Photolysis of protoporphyrin-treated human fibroblasts in vitro: Studies on the mechanism. J. Lab. Clin. Med. 96, 158-167.

34. Weishaupt, K.R., Boyle, D.C., Mensinger, M. and Dougherty, T.J. (1979) Cellular and tissue repair of photodynamic damage. Proc. Am. Assn. Cancer Res. 20, 918.

HEMATOPORPHYRIN DERIVATIVE: CHEMICAL COMPOSITION, PHOTOCHEMICAL AND PHOTOSENSITIZING PROPERTIES

Johan Moan*†, Sverre Sandberg‡, Terje Christensen† and Stein Elander†**

†Norsk Hydro's Institute for Cancer Research,
Montebello
Oslo 3, Norway

‡Laboratory of Clinical Biochemistry
University of Bergen
5016 Haukeland
Sykehus, Norway

**Institute of Chemistry
University of Oslo
Oslo, Norway

*To whom correspondence should be addressed.

INTRODUCTION

"Hematoporphyrin derivative" (HPD) is a product prepared by treating acetylated hematoporphyrin (HPA) with alkali. Acetylated hematoporphyrin is prepared by treating hematoporphyrin dihydrochloride with a mixture of sulphuric acid and acetic acid (Lipson et al. 1961). HPD has been reported by a number of investigators to be preferentially retained in tumors (for a review, see Moan and Christensen, 1980). HPD is now being evaluated for use in cancer detection and therapy. The therapeutic aspect is based on the fact that many porphyrins are efficient photosensitizers. Tumor tissue, which after injection contains more HPD than surrounding tissue, is selectively destroyed when exposed to visible light (Dougherty et al., 1975).

Other investigators have shown HPA as well as HPD to be mixtures of porphyrins (Clezy et al., 1980, Bonnett et al., 1980). Different porphyrins have different photosensitizing properties (Kessel, 1977; Sandberg and Romslo, 1981), and it would therefore be of considerable interest to separate the components of HPD, to identify them chemically and to determine their individual properties as chromophores, photosensitizers and tumor-seekers.

MATERIALS AND METHODS

Acetylation of hematoporphyrin was carried out by treating hematoporphyrin dihydrochloride (Koch-Light) with a mixture of sulphuric and acetic acids as described by Lipson et al. (1961). This product will be termed acetylated hematoporphyrin or HPA. Alkali treatment of HPA was carried out by dissolving 6.3 mg/ml HPA in 0.1 M NaOH and neutralizing the solution one hour later to pH 7.3 by dropwise addition of 0.1 M HCl. In correspondence with the major trend in the literature describing the tumor localizing properties of porphyrins, the alkali-treated HPA will be termed hematoporphyrin derivative or HPD. The stock solution was kept at -20°C and was stable for more than half a year. Hematoporphyrin free base and tryptophan were obtained from Sigma. Protoporphyrin and 2(4)-hydroxyethyl-4(2)vinyl-deutero porphyrin were obtained from Porphyrin Products, Logan, Utah. [^{3}H]-labelled acetic acid was obtained from Amersham International Ltd., England.

Absorption and fluorescence spectra were recorded with a Cary 118 spectrophotometer and a Hitachi Perkin-Elmer (MPF-2A) spectrofluorometer, respectively. Quantum yields of fluorescence at room temperature were determined by comparison with a standard: fluorescein in 0.1 M NaH_2CO_3, pH = 9.6, quantum yield = 0.85 (Parker, 1968). Sample and fluorescein standard with the same OD at the excitation wavelength (366 nm) were exposed to the light from a HBO 200 W high pressure mercury lamp fitted to a Bausch & Lomb grating monochromator. The fluorescence intensity was measured with a calibrated photodetector (UDT 11A, United Detector Technology, Santa Monica, CA, U.S.A.). The accuracy of the method was estimated by measuring the fluorescence quantum yields of other standards: anthracene, quinine sulfate, proflavin and eosin. In no case did the determined quantum yield deviate more than 20% from accepted values from the literature (Parker, 1968).

Singlet oxygen yields were determined by measuring the sensitized photodegradation of tryptophan as described elsewhere (Kvello-Stenström et al., 1980).

HPLC analysis was carried out with a liquid chromato-

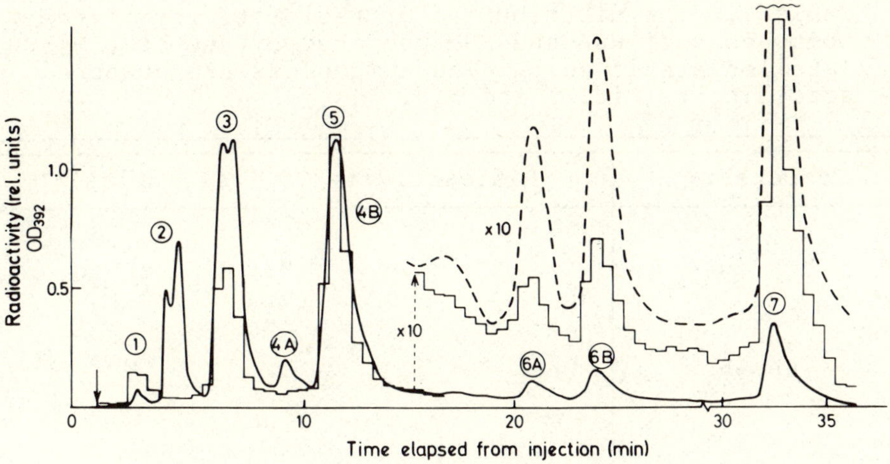

Figure 1. Chromatogram of acetylated hematoporphyrin (continuous curve) and distribution of radioactive acetate groups (histogram). Column: Supelcosil LC-18 (250x4.6 mm). Mobile phase: 0-28 min: Methanol/water, in volume ratio 82:18, 1 mM phosphate, pH=5.7; 28-40 min: Methanol/water in volume ratio 82:18, 1 mM phosphate, pH 8.5. Flow rate: 2 ml/min. Detection by optical absorption at 392 nm. Radioactivity measured in fractions of 1.0 ml.

graph equipped with a variable wavelength detector (LDC spectro-monitor III, model 1204) set at 392 nm. A Supelcosil LC 18 column (250 x 4.6 mm) was used for the analytical as well as for the semipreparative part of the work. The mobile phase was methanol and water, usually in volume ratio 82:18 and containing 1 mM sodium phosphate. The flow rate was 2 ml/min. Fractions containing the various components of HPD were collected, the solvent was evaporated under reduced pressure and the samples were stored at -20°C in evacuated tubes.

To identify the components that contained acetate groups, radioactively-labelled [^{3}H] acetic acid was used for preparation of HPA.

HPA and HPD, as well as the four main components of HPD, were subjected to esterification and analysed in a normal phase chromatographic system. The porphyrins were esterified by treatment with 10% sulphuric acid in methanol for 18 hr in the dark. The solutions were brought to pH 3.5 by addition of sodium acetate, the esters were extracted into chloroform and washed with distilled water. Samples of

Table 1. The distribution of radioactivity in the components of HPA and HPD prepared by use of [^{3}H]-labelled acetic acid. The components are numbered according to Fig. 1 and Fig. 3.

Component*	Radioactivity/OD (rel. units)**
2	0
3	0.51
4A,4B	0
5	1.00
6A	0.55
6B	0.47
7	0.50
7,alkali-treated	0.064

* Component 1 (Fig. 1) is not included since it is eluted simultaneously with acetate groups dissociated from the porphyrin ring.

** In replicate experiments the values were reproduced within 15% of the numbers given.

the chloroform phase were injected in a μPorasil column (300 x 4 mm) with a Waters Model 6000 Solvent delivery system, a U6K injector and a Model 440 absorbance detector (filter 405 nm). The mobile phase was n-heptane/methyl acetate (volume ratio 3:2) and the flow rate was 1.5 ml/min.

RESULTS

Fig. 1 and Fig. 2 show chromatograms of HPA together with the distribution of the radioactive label. More than 95% of the injected material was recovered from the columns in runs as shown in Fig. 1. The ratio of the radioactivity and the integrated optical absorption of the fractions containing the separate components are given in Table 1. When used in animal and clinical experiments HPD is treated with

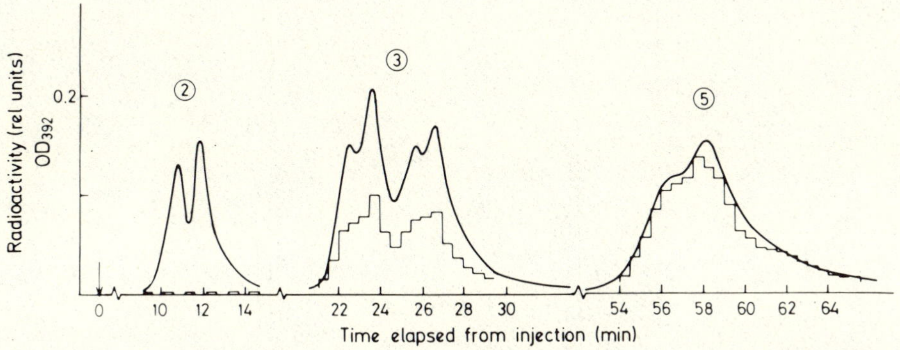

Figure 2. Chromatograms and radioactivity of components 2, 3 and 5 (see Fig. 1) under conditions described for Fig. 1 except methanol/water ratio 74/26 (pH 5.7).

0.1 M NaOH for 1 hr and then neutralized by addition of HCl before injection (Lipson et al., 1961, Dougherty et al., 1979). This caused a marked change in the chromatogram (Fig. 3). All radioactive acetate was found at the solvent front (data not shown), except for a small fraction that remained associated with component 7 (Table 1). The stabilities of components 3 and 5 in phosphate buffered saline (PBS) at pH 7.3 were tested separately. At room temperature the half-lives were 10 and 50 minutes, respectively. The chromatograms of 2(4)hydroxyethyl-4(2)-vinyl deuteroporphyrin (HVD) and protoporphyrin are also shown in Fig. 3. Porphyrins are often analysed in normal phase chromatographic systems after being converted to their methylesters. Fig. 4 shows the chromatograms of esterified HPA and HPD in a chromatographic system with a polar stationary phase. Components 2, 4A and 4B from the reversed phase system (Fig. 3) were esterified and chromatographed separately and found to correspond to the peaks marked 2, 4A and 4B on Fig. 4. Components 4A and 4B changed their relative positions compared with Fig. 3. The peak marked PP had a retention time identical with that of protoporphyrin methyl ester.

Aggregation of porphyrins in aqueous solutions may be brought about by increasing their concentration or by lowering the pH. This is shown by Fig. 5 which compares the Soret band absorption spectrum of hematoporphyrin at different concentrations and pH values. Monomers have an absorption maximum at about 390 nm, while the maximum for aggregates is 370 nm. When OD_{390}/OD_{370} is unity, somewhat more than half of the porphyrin amount is present as aggregates. (The optical absorption per porphyrin ring is reduced upon dimerization and/or aggregation.) For the com-

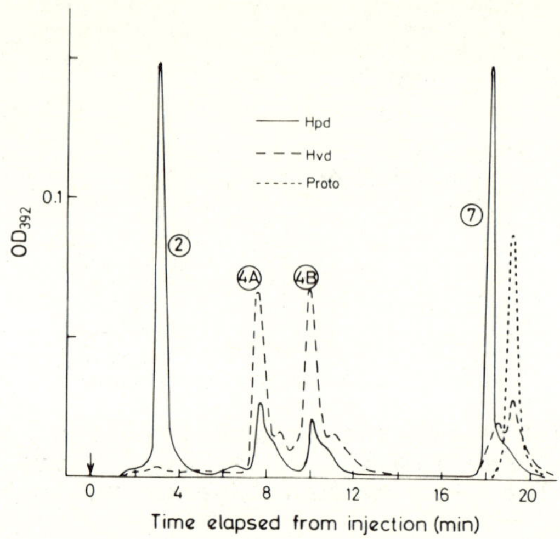

Figure 3. Chromatograms of protoporphyrin (...), 2(4)hydroxy-ethyl-4(2)-vinyldeuteroporphyrin (---) and alkali treated HPD (——) under conditions described in the legend to Fig. 1 except mobile phase from 12 min was methanol/water in volume ratio 74:26 (pH 8.5).

ponents 2, 4A, 4B and 7, OD_{390}/OD_{370} was unity at the following values of OD_{390}: 1.6 (approx. 1.6 x 10^{-5} M), 0.1, 0.1 and 10^{-3}, respectively.

Fig. 6 shows the titration curve for hematoporphyrin, the fluorescence quantum yield, wavelength absorption maximum and the yield of singlet oxygen as functions of pH. The quantum yield of fluorescence is much higher for monomers than for aggregates. This is illustrated by Fig. 7 which shows the absorption spectrum and the fluorescence excitation spectrum of component 4A of HPD. Similar spectra were obtained for all the components of HPD.

Table 2 shows quantum yields of fluorescence and singlet oxygen measured at pH 7.3 in PBS and in 10 mM phosphate buffer in D_2O, respectively. D_2O was used in the latter measurements since the singlet oxygen lifetime is 10 X longer in D_2O than in H_2O (Gorman and Rodgers, 1978).

DISCUSSION

In the process of preparing HPA from hematoporphyrin dihydrochloride all the side groups are expected to be

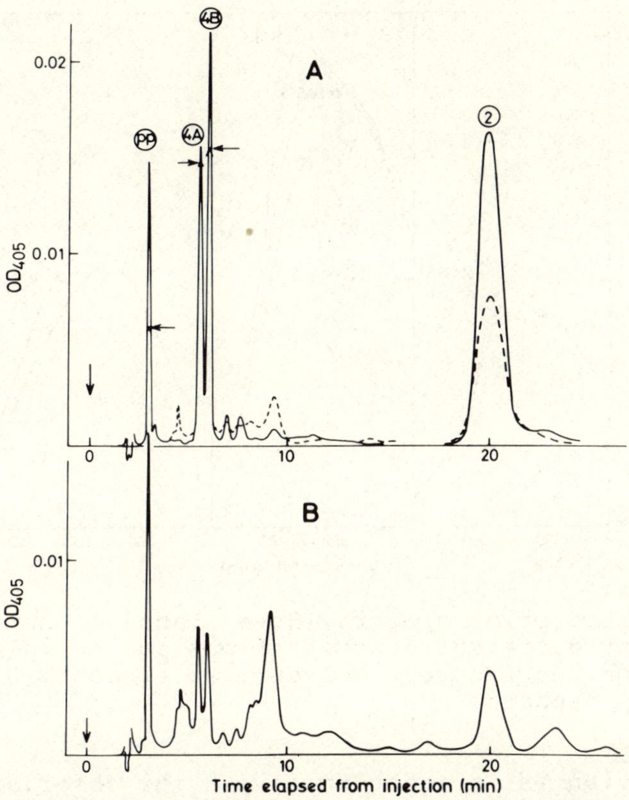

Figure 4. The upper part (A) of the figure shows chromatograms of esterified HPA (——) and HPD (---) in a normal phase system. The lower part of the figure (B) shows the chromatogram of esterified component 7, from HPD (see Fig. 3) under similar conditions. Column: μPorasil (300 x 4 mm). Mobile phase: n-heptane/methyl acetate in volume ratio 3/2. Flow rate: 1.5 ml.min. Detection by optical absorption at 405 nm. The horizontal arrows indicate the heights of the peaks of the chromatogram of HPD.

stable except those at positions 2 and 4 (Fig. 8) (Clezy et al., 1980). The following groups are likely to be present at these positions in order of decreasing polarity: CH(OH)Me, CH(OAc)Me and $CH=CH_2$ (Clezy et al., 1980). Our data (Figs. 1 and 2, Table 1) indicate that two acetate groups are present in component 5, while one acetate group is present in each of the components 3A, 3B, 6A and 6B. A tentative identification of the components was made (Table 4). Our semipreparative system does not have capacity to yield samples large enough for NMR spectroscopy. But our

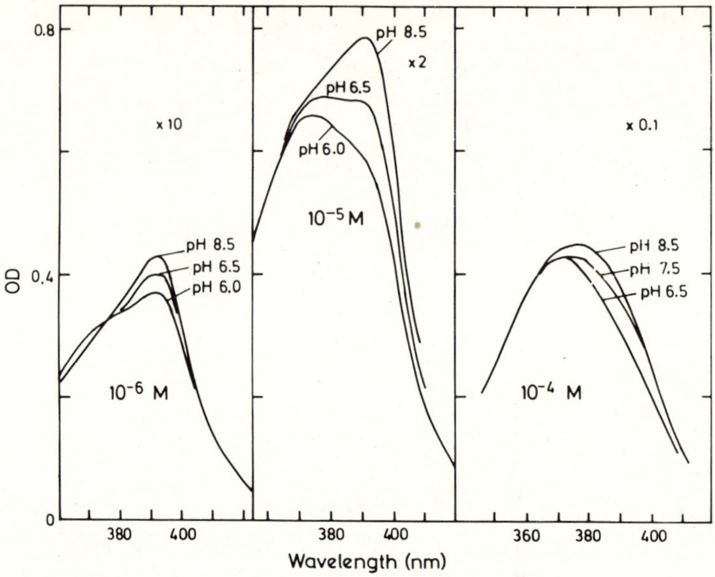

Figure 5. Absorption spectra (Soret band) of hematoporphyrin at different concentrations and pH values as shown on the figure. Solvent: water containing 10 mM phosphate.

identification is in accordance with the observations that the chromatogram of hematoporphyrin (Porphyrin Products) corresponds to component 2 (in both chromatographic systems) and that the chromatogram of HVD corresponds to component 4. Furthermore, our assignments are in accordance with those of Bonnett et al. (1980), with one exception. Bonnett identified a component with long retention time on a reversed phase column as protoporphyrin. Fig. 4A indicates that some protoporphyrin may be present in component 7, but Fig. 3 shows that protoporphyrin has a longer retention time than the main part of component 7. We also observed protoporphyrin to be less water soluble than component 7 (results not shown).

The least polar component of HPA has the same retention time as the least polar component of HPD, hence it has been given the same assignment (i.e. 7). However, Table 1 clearly shows that component 7 of HPA contains much more radioactivity than component 7 of HPD. Therefore, we may assume that these components are chemically different. It is of interest to note that the acetate groups of hematoporphyrin mono- and di-acetate hydrolyze completely during the treatment with alkali, while some acetate groups evidently remain associated with component 7 after alkali treatment.

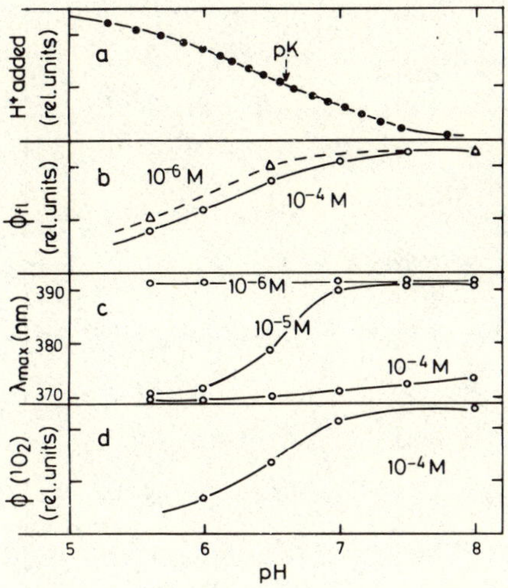

Figure 6. Data on hematoporphyrin in aqueous solution. Part a) Titration curve of a 0.1 mM solution. Part b) Fluorescence in aqueous solutions at different pH values. Part c) Wavelength of absorption maximum at different concentrations and pH values. Part d) Singlet oxygen quantum yield as a function of pH.

Fig. 4B shows that esterified component 7 of HPD contains a number of porphyrins, among them evidently the esters of hematoporphyrin and 2(4)hydroxyethyl-4(2)-vinyldeuteroporphyrin. This may indicate that component 7 is unstable during the esterification and extraction procedure. However, in view of the work of Dougherty et al. and Bonnett et al. presented in this book, it is more likely that component 7 contains aggregates of several porphyrins, and that these aggregates are monomerized in the normal phase chromatographic system. Aggregation may also partly explain the presence of radioactivity in component 7 (Fig. 1). Further work is needed to identify this component.

Our chromatographic system not only separates positional isomers, but also separates diastereomers. This is indicated by Fig. 2 which shows that radioactive acetate groups are present in equal amounts in the subfractions of component 3 and component 5. Thus, it seems that LL and DD forms may be separated from LD and DL forms. A similar separation of bacteriochlorophyll-c was reported by Smith et al. (1980) who used a similar chromatographic system.

Alkali treatment of HPA resulted in a marked change in its chromatogram in the reversed phase system (Figs. 1 and

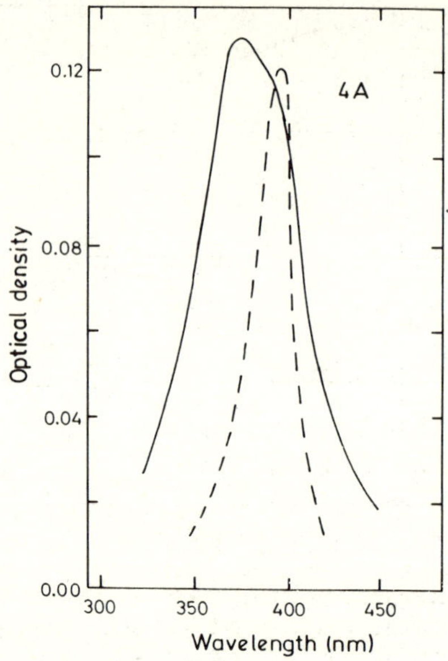

Figure 7. Absorption spectrum of component 4A of HPD in aqueous solution at pH 7.3. The dotted curve shows the fluorescence excitation spectrum recorded under identical conditions.

3). The chromatograms of the corresponding methyl esters in a normal phase system were almost identical (Fig. 4). This indicates that components 3 and 5 were unstable during the esterification, the washing or the extraction process.

Alkali treatment of HPA obviously resulted in hydrolysis of the acetates. The components containing acetate groups are unstable even in neutral aqueous solution. Thus, neither hematoporphyrin diacetate nor hematoporphyrin monoacetate (components 3 and 5) seem to be important for the tumor-localizing properties of HPD. As shown by Table 3, component 7 is the most efficient photosensitizer. The last column of this table shows that the photosensitizing properties of HPD are mainly due to this component.

It is well known that porphyrins tend to dimerize or aggregate in aqueous solution (White, 1978). Drug dimers are often difficult to distinguish from higher aggregates; in the present work we have not made such distinctions. However, it has been proposed that porphyrin dimers may be the predominant aggregate form under conditions used in the present study (Gallagher and Elliot, 1973).

Table 2. Fluorescence quantum yields (ϕ_{fl}) in absolute units and singlet oxygen quantum yields $\phi(^1O_2)$ in relative units. Experimental conditions described in the text. The optical density at 366 nm was 0.2 for all components.

Component	ϕ_{fl}*	$\phi(^1O_2$*)
2	0.071	2.4
4A	0.043	1.2
4B	0.039	1.1
7	0.0036	0.2

* In replicate studies the numbers were reproducible to within 20% of the values shown.

Table 3. Composition of HPD and the relative efficiencies of its components in sensitizing NHIK 3025 cells to photoinactivation.

Component	% Total	Rel.sensitizing efficiency (E)	$A_{360}E$
2	27	0.07	1.9
4A	18	0.30	5.4
4B	16	0.44	7.0
7	34	1.0	34

The % Total was estimated by measuring the optical absorption at 360 nm. The sum of the optical densities of the components was 96% of the optical density of the injected amount of HPD. Relative sensitizing efficiencies are taken as $t(10)^{-1}$, where t(10) is the exposure time needed to inactivate 90 per cent of the cells after incubation for 1 hr with HPD components at 37°C at a concentration corresponding to OD_{360} = 0.2.

Table 4. Chemical structure of the components of HPA and HPD (Cf. Figs. 1-3,8).

Component*	R^2	R^4	N_C**	Stability in water at pH 7***
2	CH(OH)Me	CH(OH)Me	2	+
3A	CH(OAc)Me	CH(OH)Me	2	-
3B	CH(OH)Me	CH(OAc)Me	2	-
4A	CH(OH)Me	$CH=CH_2$	1	+
4B	$CH=CH_2$	CH(OH)Me	1	+
5	CH(OAc)Me	CH(OAc)Me	2	-
6A	$CH=CH_2$	CH(OAc)Me	1	-
6B	CH(OAc)Me	$CH=CH_2$	1	-
7HPA				-
7HPD				+
Protoporphyrin	$CH=CH_2$	$CH=CH_2$	0	+

*A and B indicate positional isomers.

**N_C = number of positions with asymmetric carbon atoms.

***Negative and positive signs denote that the half life of the compound is less or more than 1 h, respectively.

The tendency of the HPD components to aggregate in aqueous solutions increases with decreasing polarity (Table 2). Furthermore, protonated neutral forms have a more pronounced tendency to aggregate than do deprotonated forms

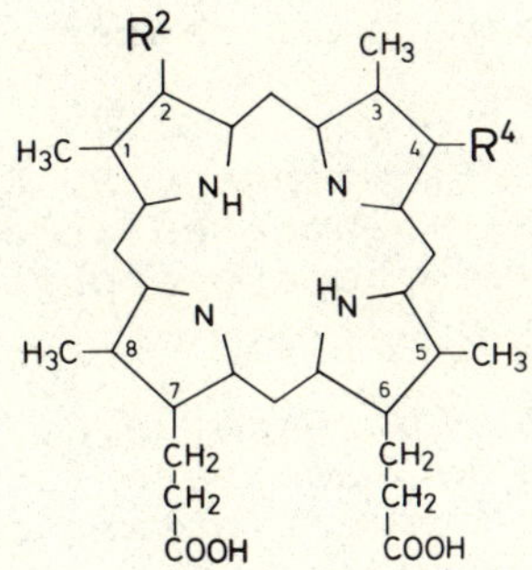

Figure 8. Chemical structure of the components of HPD. See Table 4.

(Fig. 5). Aggregation quenches the fluorescence (Table 2 and Fig. 7) and diminishes singlet oxygen production (Table 2). Since singlet oxygen is produced via the triplet state, and the fluorescence and singlet oxygen quantum yields are quenched by nearly the same factor (Table 2), quenching may occur before intersystem crossing to the triplet state. In solutions of low porphyrin concentration, protonation may occur upon lowering pH without any accompanying aggregation (Fig. 5 left, curves). Data shown in Fig. 6b suggest that aggregation and protonation result in fluorescence quenching.

ACKNOWLEDGEMENTS

The authors want to thank Tyge Greibrokk, Jorolf Alstad and Jan Karlsen for valuable comments. This study was supported by the Norwegian Cancer Society and the Norwegian Research Council for Science and the Humanities.

REFERENCES

1. Bonnett, R., R.J. Ridge and P.A. Scourides (1980) Haematoporphyrin derivative. J.C.S. Chem. Comm. 1198-1199.

2. Clezy, D.S., T.T. Hai, R.W. Henderson and L. van Thuc (1980) The chemistry of pyrrolic compounds. XLV. Haematoporphyrin derivative: Haematoporphyrin diacetate as the main product of the reaction of haematoporphyrin with a mixture of acetic and sulphuric acids. Aust. J. Chem. 33, 585-597.

3. Dougherty, T.J. (1980) Hematoporphyrin derivative for

detection and treatment of cancer. J. Surg. Oncol. 15, 209-210.

4. Dougherty, T.J., G.B. Grindey, R. Fiel, K.R. Weishaupt and D.G. Boyle (1975) Photoradiation therapy. II. Cure of animal tumors with hematoporphyrin and light. J.Natl. Cancer Inst. 55, 115-121.

5. Dougherty, T.J., G. Lawrence, J.H. Kaufman, D. Boyle, K.R. Weishaupt and A. Goldfarb (1979) Photoradiation in the treatment of recurrent breast carcinoma. J. Natl. Cancer Inst. 62, 231-237.

6. Gallagher, W.A. and W.B. Elliot (1973) Ligand binding in porphyrin systems. Ann. N.Y. Acad. Sci. 206, 463-482.

7. Gorman, A.A. and M.A.J. Rodgers (1978) Lifetime and reactivity of singlet oxygen in an aqueous micellar system: a pulsed nitrogen laser study. Chem. Phys. Lett. 55, 52-54.

8. Kessel, D. (1977) Effects of photoactivated porphyrins at the cell surface of leukemia L1210 cells. Biochem. 16, 3443-3449.

9. Kvello-Stenstrøm, A.G., J. Moan, G. Brunborg and T. Eklund (1980) Photodynamic inactivation of yeast cells sensitized by hematoporphyrin. Photochem. Photobiol. 32, 349-352.

10. Lipson, R.L., E.W. Baldes, A.M. Olsen (1961) The use of a derivative of hematoporphyrin in tumor detection. J. Natl. Cancer Inst. 26, 1.

11. Moan, J. and T. Christensen (1980) Porphyrins as tumor localizing agents and their possible use in photochemotherapy of cancer. Tumor Res. 15, 1-10.

12. Moan, J., L. Smedshammer and T. Christensen (1980) Photodynamic effects on human cells exposed to light in the presence of hematoporphyrin. pH effects. Cancer Lett. 9, 327-332.

13. Parker, C.A. Photoluminescence of solutions, pp. 266-267 London, Amst., New York (1968).

14. Sandberg, S. and I. Romslo (1981) Porphyrin-induced photodamage at the cellular and the subcellular level as related to the solubility of the porphyrin. Clin. Chim. Acta 109, 193-201.

15. Smith, K.M., G.E.F. Bisset and M.J. Bushell (1980) Recent studies on the chlorobium chlorophylls

(Bacteriochlorophyll-c) Int. J. Biochem. 12, 695-700.

16. White, W.I. Aggregation of porphyrins and Metalloporphyrins. In: The Porphyrins. Ed.: D. Dolphin. pp. 303-336. Academic Press. New York, San Francisco, London, 1978.

A PRELIMINARY COMPARISON OF THE PHOTOSENSITIZING PROPERTIES OF PORPHYRINS IN AQUEOUS SOLUTION AND LIPOSOMAL SYSTEMS

John D. Spikes

Department of Biology
University of Utah
Salt Lake City, Utah 84112

INTRODUCTION

Most of the detailed mechanistic studies on porphyrins as photosensitizers have been carried out in more or less homogeneous solution. Sensitized photoreactions in cells, as exemplified by the porphyrin-sensitized photokilling of tumor cells, must be a much more complex process. Cells are non-homogeneous with abrupt phase changes and with major differences in chemical composition in different regions. Because of localized differences, sensitizers tend to selectively dissolve in or bind non-covalently to different micro-regions of the cell; on illumination this could result in selective, localized damage to cell structure and function. Because of these complexities, we know relatively little about the specific lesions in photosensitized cells leading to injury and death (1).

A sometimes useful approach to biological complexity is the use of model systems for experimental studies, models which simulate, although in a very oversimplified way, some of the properties of living cells. Several types of model systems are presently being used in photosensitization studies. One consists of sensitizer bound to protein molecules in aqueous solution; this system simulates in some respects sensitizer molecules associated with macromolecules in cells (1). Surfactant micelles in aqueous media have both polar and non-polar regions and are being used as membrane models in photosensitization studies (1). Finally, liposomes (rather stable, microscopic, bilayered phospholipid vesicles), which have been used extensively as models for cell membranes (2), are beginning to be used in photosensitization studies (3-7).

This paper will summarize some of our recent preliminary work on the sensitizing behavior of "water-soluble" porphyrins (especially hematoporphyrin) in aqueous solution and entrapped in aqueous solution within liposomes as well as lipid-soluble porphyrins (in particular, hematoporphyrin dimethyl ester) incorporated into liposomal membranes. The absorption, emission, photobleaching and flash photolysis behavior of porphyrins in these systems will be described as well as the photodynamic effects of porphyrins in these systems on small biomolecules (unsaturated phospholipids, amino acids, etc.), proteins and mammalian cells in culture.

MATERIALS AND METHODS

The porphyrins used were from Porphyrin Products, Inc., the lysozyme from the Worthington Biochemical Corp., and the lecithins (phosphatidyl cholines) from Sigma Chemical Co. All of the other chemicals were of the highest purity available. Absorption spectra were recorded with a Cary model 14 spectrophotometer and fluorescence emission spectra (uncorrected) were measured with an Aminco-Bowman spectrofluorimeter.

Small, unilamellar vesicles (SUV) were used for the liposome experiments (2). The phospholipid (plus a lipid-soluble porphyrin when desired) was dissolved in chloroform:methanol 9:1, the solvent was evaporated off, the lipid film was resuspended in aqueous buffer, and the suspension was sonicated under flowing nitrogen for 30 min. The resulting liposomes were unilamellar, of fairly even size, and had an average diameter of ~40 nm. For liposomes with entrapped sensitizers, the lipid film was resuspended and sonicated in aqueous buffer containing the dissolved sensitizer; external sensitizer was removed using Sephadex G-25. Liposomes were prepared from egg phosphatidyl choline (EPC, approximately half of the side chains unsaturated, phase transition temperature -15 to -7°C) and dipalmitoyl phosphatidyl choline (DPPC, side chains saturated, transition temperature 41°C).

Flash experiments were carried out with a Phase-R DL-2100C dye laser with 60 μM coumarin 504 (499nm) in methanol. The apparatus and the computer-aided control, data analysis system and routines are described elsewhere (8). Quantum yields of oxygen uptake during the sensitized photooxidation of substrates were measured in air at 25°C with a recording membrane-type oxygen electrode system. Illumination was provided by a 500 W slide projector and multilayer interference filters (9). The quantum yields of lysozyme photoinactivation were measured as described elsewhere (9,10).

For photodynamic killing experiments, approximately 10^5 HeLa cells (strain D98/AH2, kindly supplied by Dr. M.C. Rechsteiner of this department) were placed in 25 cm^2 plastic culture flasks and incubated for 24 hr to permit good attachment. Cells were illuminated in air in Ham's F-12 medium (without serum) containing dissolved porphyrins or porphyrin-loaded liposomes. After rinsing, cells were cultured in complete medium; at 4 and 24 hr, counts were made of live and dead cells using a phase contrast microscope. Flasks were illuminated from the bottom using a 500 W slide projector with the broad band incident light intensity adjusted to 5×10^5 erg cm^{-2} sec^{-1}. A shielding adaptor permitted illuminating four 1.2 x 2.5 cm areas of the flask for different times; a fifth area was protected from light and served as a dark control. Thus a four-dose survival curve was obtained from each flask.

RESULTS AND DISCUSSION

Absorption, Emission, Photostability and Flash Photolysis Behavior of Porphyrins in Solution and in Liposomes

The Soret peak of hematoporphyrin in 0.1 M pH 7.4 phosphate buffer progressively shifted from ~390 nm at low concentration (1.25μM) to ~373 nm at high concentration (400 μM). For 12.5 μM hematoporphyrin in 0.1 M phosphate buffer, the peak shifted from ~389 nm at pH 10.5 to ~373 nm at pH 5.0. These changes apparently result from an increase in hematoporphyrin aggregation with increasing concentration and with decreasing pH. If Triton X-100 was added to hematoporphyrin solutions in buffer to a concentration of 1%, the absorption peak became narrower and higher, and the maximum shifted to 400 nm; this suggests that only monomer porphyrin is present in detergent solutions. The absorption peak of hematoporphyrin dissolved in chloroform:methanol 9:1 was at 400 nm; the porphyrin would be expected to be in monomeric form in this solvent. These data all suggest, as other investigators have noted (11), that hematoporphyrin in aqueous media is present as a mixture of monomers and various aggregates. If a 400 μM solution of hematoporphyrin in 0.1 M pH 7.4 phosphate buffer (absorption maximum at 374 nm) was entrapped inside DPPC liposomes, the spectrum of the suspension peaked at 400 nm. This probably results from the diffusion of porphyrin from the aqueous medium into the highly non-polar core of the liposomal membrane where it can exist as the monomer.

Hematoporphyrin dimethyl ester dissolved in chloroform:methanol 9:1 had the Soret peak at 400 nm over a wide concentration range suggesting that the porphyrin was present in this solvent as the monomer. Incorporation of the ester into the membranes of liposomes did not change the peak indicating that the porphyrin was in monomeric form in this environment. This was true in liposomal membranes with porphyrin ester:phospholipid molecular ratios ranging from 1:1400 to 1:17.5. The absorption peaks of these preparations did not change on the addition of Triton X-100, which destroys the ordered liposome structure. Hematoporphyrin diamyl ester and dipalmitoyl ester behaved in the same way. However, not all lipid-soluble porphyrins incorporated into liposomal membranes as the monomer. For example, octaethyl porphine had a very sharp Soret peak at ~399 nm as a solution in chloroform:methanol; when the porphyrin was incorporated into liposomal membranes, the absorption peak became low and broad, very much like the absorption spectra of colloidal porphyrin esters suspended in aqueous buffers.

Hematoporphyrin in 0.1 M pH 7.4 buffer showed a sharp fluorescence emission peak at ~611 nm; the expected second emission peak appeared only as a shoulder on the long wavelength side of the first peak because of the low sensitivity in the red region of the photomultiplier tube used in the fluorimeter. The emission peak of hematoporphyrin dimethyl ester in chloroform:methanol was at ~625 nm while the peak for the ester incorporated into the membranes of DPPC liposomes was at ~627 nm; again, this suggests that the ester in the membrane is in the same monomeric form as in the organic solvent.

Most porphyrins can be converted into colorless products (photobleached) on exposure to high intensity light. Table 1 shows data on the photobleaching of hematoporphyrin and hematoporphyrin dimethyl ester as measured in different environments. Hematoporphyrin in buffer was relatively unstable on illumination, while the ester dissolved in chloroform:methanol was very stable. The ester was somewhat less photostable when incorporated into the membranes of DPPC liposomes, and much less stable when incorporated into liposomes prepared from the unsaturated phospholipid.

A few preliminary flash experiments were carried out to compare the properties of porphyrin triplet states in aqueous solution and in phospholipid membranes. Solutions 80 μM in hematoporphyrin in 0.1 M pH 7.4 phosphate buffer showed an apparent (uncorrected) triplet peak at ~430 nm. Under anaerobic conditions the triplet decay pattern was best fit as competing first and second order reactions; The latter process probably represents triplet-triplet quenching interactions. The rate constant for the first order component of the decay was 1×10^4 sec^{-1}. The apparent triplet

Table 1. Photobleaching of porphyrins: percent of peak absorbance remaining after 60 min of illumination in air at 25°c with a light intensity (incandescent lamps) of 7.2×10^6 erg cm^{-2} sec^{-1}. The liposomes had a porphyrin:phospholipid molecular ratio of 1:70

Porphyrin System	% Absorbance
5 μM Hematoporphyrin in 0.1 M phosphate buffer, pH 7.4	59
5 μM Hematoporphyrin dimethyl ester in chloroform:methanol 9:1	95
Hematoporphyrin dimethyl ester in DPPC liposomes	70
Hematoporphyrin dimethyl ester in EPC liposomes	33

peak for hematoporphyrin dimethyl ester incorporated into saturated (DPPC) liposomes with an ester:phospholipid molecular ratio of 1:70 was at ~470 nm. The decay of this transient was almost entirely first order with a rate constant of 2.2×10^3 sec^{-1}. Thus the lifetime of the triplet state of hematoporphyrin dimethyl ester in DPPC liposomes is several times greater than that of hematoporphyrin in aqueous buffer, as measured under the conditions indicated.

Porphyrin-Sensitized Photooxidation of Biomolecules in Aqueous Buffer and in Liposomal Systems

There is no simple way to compare the photosensitizing efficiencies of porphyrins in aqueous solution and when incorporated into liposomal membranes. Table 2 shows the effect of the pH on the quantum yield of oxygen uptake during the sensitized photooxidation of histidine by hematoporphyrin in solution and by different lipid-soluble porphyrins in DPPC liposome membranes. The quantities of liposomes added to the reaction systems were such that the "average" porphyrin concentration (as calculated assuming that it was uniformly distributed in the reaction mixture) was the same as the hematoporphyrin concentration in solution. As may be seen, hematoporphyrin dimethyl ester in liposomes was the most efficient sensitizer over the entire pH range. The observed pH dependence reflects the fact that only the neutral form of the histidine molecule is photooxidizable; thus the yield follows the ionization curve (12). The yield was increased 2-3 fold in D_2O and was inhibited 50% by 8×10^{-4} M sodium azide. This suggests that singlet oxygen is involved in the photooxidation; histidine is known to react very efficiently with singlet oxygen (12).

Hematoporphyrin in solution was quite an efficient sensitizer for the photooxidation of histidine at high pH, but the efficiency decreased more sharply with decreasing pH than in the case of the ester. This probably results from the increase in aggregation of the porphyrin with decreasing pH as discussed in the preceding section. Protoporphyrin dimethyl ester in liposomes was somewhat less efficient as a sensitizer than the hematoporphyrin ester. The photosensitizing efficiency of octaethyl porphine in liposomes was very low. This is undoubtedly related to the observation described above, that this porphyrin, although very soluble in chloroform:methanol, did not incorporate into liposomal membranes as the monomer. Colloidal hematoporphyrin dimethyl ester suspended in buffer was essentially without sensitizing activity. This suggests that large aggregates of sensitizing porphyrins may not sensitize photooxidations efficiently.

The photooxidation of liposomes prepared from an unsaturated phospholipid (EPC) as sensitized by membrane-

Table 2. Quantum yields for oxygen uptake during the porphyrin sensitized photooxidation of histidine in air at 25°C. The reaction mixtures were 0.1 M in phosphate buffer and 2 mM in histidine. The hematoporphyrin concentration and the "average" porphyrin concentration in the liposome experiments were 12.5 μM. The liposomes had a porphyrin:phospholipid molecular ratio of 1:70.

Porphyrin System	pH 5	6	7	8	9	10
Hematoporphyrin dihydrochloride	0	0.008	0.044	0.075	0.10	0.12
Hematoporphyrin dimethyl ester in DPPC liposomes	0.009	0.042	0.076	0.10	0.12	0.13
Protoporphyrin dimethyl ester in DPPC liposomes	0.004	0.022	0.046	0.066	0.082	0.096
Octaethyl porphine in DPPC liposomes	0	0	0.001	0.001	0.002	0.002

Table 3. Quantum yields (X 10^3) for the porphyrin-sensitized photodynamic inactivation of lysozyme as a function of pH as measured in air at 25°C. The lysozyme concentration was 100 µM and the reaction medium was 0.1 M phosphate buffer. The hematoporphyrin concentration and the "average" porphyrin concentration in the liposome experiment were 25 µM. The liposomes had a porphyrin:phospholipid molecular ratio of 1:70.

Porphyrin System	pH 5	6	7	8	9	10
Hematoporphyrin dihydrochloride	0.8	3.2	5.2	6.1	6.8	7.2
Hematoporphyrin dimethyl ester in DPPC liposomes	2.9	4.0	4.5	4.7	4.8	4.9

incorporated hematoporphyrin dimethyl ester and external hematoporphyrin in solution was then examined. With ester incorporated in the membrane at a porphyrin:phospholipid molecular ratio of 1:70, the quantum yield of oxygen uptake on illumination was 0.015 at pH 10; this decreased slightly with decreasing pH and reached a value of 0.012 at pH 5. With 12.5 µM hematoporphyrin in the external solution, the quantum yield of oxygen uptake was only 0.005. The yield increased slightly with decreasing pH, perhaps as a result of a more efficient penetration of the porphyrin into the liposomal membrane. Lipid soluble substrates (diphenylisobenzofuran and linolenic acid methyl ester) incorporated into the membranes of DPPC liposomes were also photooxidized with membrane-incorporated and external porphyrins; DPPC is not photooxidized. Suwa et al. (4) found that cholesterol incorporated in liposomal membranes was photooxidized to the singlet oxygen product with hematoporphyrin; membrane-incorporated porphyrin was a much more efficient sensitizer than porphyrin in the external medium.

Finally, a comparison was made of the efficiency of the photodynamic inactivation of lysozyme by hematoporphyrin in solution and hematoporphyrin dimethyl ester in DPPC liposomal membranes as shown in Table 3. At low pH, where the hematoporphyrin is in a more aggregated form, the liposome system was more efficient. However, at high pH, the porphyrin in solution was the most efficient. These experiments show that a lipid-soluble porphyrin incorporated into the membranes of liposomes is still an efficient sensitizer for external macromolecules. The photooxidation of the enzyme is probably mediated by singlet oxygen diffusing out of the

Table 4. Surviving fractions of HeLa cells illuminated in air in the presence of different porphyrin photosensitizing systems for the time period shown. Cell counts were made 24 hr after the end of the illumination period.

Porphyrin System	Light Dose (min)				
	0	0.35	0.75	1.5	3.0
Hematoporphyrin.diHCl (40 μM) in solution	0.99	0.92	0.26	0	0
Hematoporphyrin.diHCl (400 μM) entrapped in DPPC-liposomes	1.0	0.99	0.90	0.79	0.08
Hematoporphyrin dimethyl ester in DPPC liposome membranes	1.0	0.98	0.85	0.46	0

membrane. Nothing is known of the interaction of lysozyme with porphyrin-containing liposomes, i.e., whether the protein binds to the liposomes or penetrates into the membrane.

Photodynamic Killing of HeLa Cells Using Porphyrins in Solution and in Liposomes

There is a very large literature on the photodynamic killing of mammalian cells in culture as sensitized by a wide variety of porphyrins (see refs. 13-16 for an introduction). In most cases the cells were sensitized using porphyrins in solution. Table 4 shows the effect of time of illumination on the photodynamic killing of HeLa cells as sensitized by hematoporphyrin in aqueous solution, by hematoporphyrin in aqueous solution entrapped in DPPC liposomes, and by hematoporphyrin dimethyl ester incorporated in the membranes of DPPC liposomes. The data in the three cases cannot be compared directly, since the effective sensitizer concentrations are undoubtedly not the same and the absorption spectra are different (the same incandescent light source was used for all experiments). However, the data do show quite clearly that both a liposome-entrapped water soluble porphyrin and a lipid soluble porphyrin incorporated in liposome membranes are effective photodynamic sensitizers for HeLa cells.

Cell killing, with a given light dose, increased as the length of time the cells were incubated with hematoporphyrin in solution or with hematoporphyrin dimethyl ester loaded liposomes was increased. If cells were incubated in hematoporphyrin solution for a given period, then rinsed, and incubated in Ham's medium (without serum), the photosensitivity decreased progressively with time as the sensitizer leaked out of the cells. In similar experiments with liposomes loaded with hematoporphyrin dimethyl ester, no decrease in sensitivity occurred suggesting that the sensitizer remained in the cells. Preliminary fluorescent microscope studies indicated that the hematoporphyrin dimethyl ester loaded DPPC liposomes went essentially quantitatively into the lysosomes of the HeLa cells, as has been shown for other cells with certain porphyrins in solution (17).

SUMMARY

The preliminary studies reported here show that at least some lipid soluble porphyrins can easily be incorporated into the membranes of liposomes prepared with different phospholipids; such porphyrins do not leach out of the membranes when the liposomes are stored in aqueous media. In this environment, the lipid soluble porphyrins are very effective photodynamic sensitizers for lipophilic substrates in the liposome membrane, for external amino acids and proteins, and for mammalian cells.

Acknowledgement - This research was supported by NIH Grant GM 27921 and NIH Biomedical Research Support Grant RR07092. I thank Wendy Matis, Lisa Matsuda and Bryce Tanner for skillful technical assistance. The laser flash experiments were performed at the Center for Fast Kinetics Research, University of Texas, Austin. The Center is supported by the Biotechnology Branch of the Division of Research Resources of NIH (Grant RR00886) and by the University of Texas at Austin. I thank Dr. M.A.J. Rogers, Dr. E.L. Powers and Mr. James Lynch of the Center for their cooperation and assistance.

REFERENCES

1. G. Jori and J.D. Spikes, Photosensitized oxidations in complex biological structures, in: "Oxygen and Oxy-Radicals in Chemistry and Biology", M.A.J. Rodgers and E.L. Powers, eds., pp. 441-457, Academic Press, New York (1981).

2. B.E. Ryman and D.A. Tyrrell, Liposomes-bags of potential, Essays Biochem. 16:49-98 (1980).

3. S.M. Anderson, N.I. Krinsky, M.J. Stone and D.C. Clagett, Effect of singlet oxygen quenchers on oxidative

damage to liposomes initiated by photosensitization or by radiofrequency discharge, Photochem Photobiol. 20:65-69 (1974).

4. K. Suwa, T. Kimura and A.P. Schaap, Reaction of singlet oxygen with cholesterol in liposomal membranes. Effect of membrane fluidity on the photooxidation of cholesterol, Photochem. Photobiol. 28:469-473 (1978).

5. M. Delmelle, Retinal sensitized photodynamic damage to liposomes, Photochem. Photobiol. 28:357-360 (1978).

6. M.R. Deziel and A.W. Girotti, Photodynamic action of bilirubin on liposomes and erythrocyte membranes, J. Biol. Chem. 255:8192-8198 (1980).

7. R. Muller-Runkel, J. Blais and L.I. Grossweiner, Photodynamic damage to egg lecithin liposomes, Photochem. Photobiol. 33:683-687 (1981).

8. M.A.J. Rodgers and A.L. Bates, A laser flash kinetic spectrophotometric examination of the dynamics of singlet oxygen in unilamellar vesicles, Photochem. Photobiol., in press (1981).

9. J.D. Spikes, Selective photooxidation of thiols sensitized by aqueous suspensions of cadmium sulfide, Photochem. Photobiol. 34:549-556 (1981).

10. F. Rizzuto and J.D. Spikes, Mechanisms involved in the chemical inhibition of the eosin-sensitized photooxidation of trypsin, Rad. Environ. Biophys. 12:217-232 (1975).

11. E. Reddi, E. Rossi and G. Jori, Factors controlling the efficiency of porphyrins as photosensitizers of biological systems to damage by visible light, Med. Biol. Environ. 9:337-351 (1981).

12. C.S. Foote, Photosensitized oxidation and singlet oxygen: consequences in biological systems, in "Free Radicals in Biology", Vol. II, W.A. Pryor, ed., pp. 85-133, Academic Press, New York (1976).

13. C.J. Gomer and D.M. Smith, Photoinactivation of Chinese hamster cells by hematoporphyrin derivative and red light, Photochem. Photobiol. 32:341-348 (1980).

14. D. Kessel, Transport and binding of hematoporphyrin derivative and related porphyrins by murine leukemia L1210 cells, Cancer Res. 41:1318-1323 (1981).

15. J. Moan and T. Christensen, Photodynamic effects on

human cells exposed to light in the presence of hematoporphyrin. Localization of the active dye. Cancer Lett. 11:209-214 (1981).

16. S. Sandberg and I. Romslo, Porphyrin-induced photodamage at the cellular and the subcellular level as related to the solubility of the porphyrin, Clin. Chim. Acta 109:183-201 (1981).

17. A.C. Allison, I.A. Magnus and M.R. Young, Role of lysosomes and cell membranes in photosensitization, Nature 209:874-878 (1966).

FACTORS GOVERNING THE MECHANISM AND EFFICIENCY OF PORPHYRIN-SENSITIZED PHOTOOXIDATIONS IN HOMOGENEOUS SOLUTIONS AND ORGANIZED MEDIA

Giulio Jori, Elena Reddi, Luigi Tomio†, and Fulvio Calzavara†

Istituto di Biologia Animale
Centro C.N.R. Emocianine
Universita di Padova

†Divisione di Radioterapia
Ospedale Civile di Padova
Padova, Italy

ABBREVIATIONS USED

CTAB, cetyl trimethylammonium bromide; CTMP, 3-carbamoyl-2,2,5,5-tetramethylpyrrolidin-1-yloxy; DMPO, 5,5-dimethyl-1-pyrroline-1-oxide; His, L-histidine; HP, hematoporphyrin; HPD, hematoporphyrin derivative; HSA, human serum albumin; Met, L-methionine; NATA, N-acetyl-L-tryptophan amide; PAP, 5,10,14,20-[4-(N,N',N'',-trimethylanil-inium) porphyrin]; SDS, sodium dodecylsulphate; TAM, tryptamine; TCPP, meso-tetra-(4-carboxyphenyl)porphine; TMPH, 2,2,6,6-tetramethylpyridinium chloride; TPPS, meso-tetra-(4-sulfonatophenyl)porphine; Tri, Triton X-100; Trp, L-tryptophan.

INTRODUCTION

Porphyrin-sensitized photoreactions in biological systems usually involve the lowest excited triplet state of the dye [^{3}P] as the reactive intermediate (Spikes, 1975). For most porphyrins [^{3}P] is formed with quantum yield higher than 0.8 (Bonnett et al., 1980). It is generally assumed that in aerated media [^{3}P] is efficiently quenched by oxygen with generation of $^1O_2$ via electronic energy transfer: typical quantum yields for $^1O_2$ production range between 0.40 for tetraphenylporphines (Maillard et al, 1980) and 0.75 for HPD (Dougherty et al., 1976). However, there is increasing evidence that the interaction of [^3P] with oxygen (type II mechanism) can originate reactive species other than $^1O_2$, including $O_2^{\bullet -}$, $OH^{\bullet}$, and H_2O_2 both in homogeneous solutions and in cell systems; [^3P] may also undergo e^- transfer to or from suitable substrates [S] initiating radical processes (type I mechanism). Selected examples of porphyrin-promoted photoprocesses not involving exclusively $^1O_2$ are shown in Table 1.

The rates of the two photoprocesses can be expressed (Foote, 1976) as $v_I = k_I[^3P][S]$ and $v_{II} = k_{II}[^3P][^3O_2]$; for any given system, the competition between type I and II reaction pathways is controlled by the relative values of substrate and oxygen concentrations, as well as by the rate constants for reaction of [^3P] with [S] and O_2. The latter parameters depend on the chemical structure of [S] and on several experimental conditions, including the aggregation state of the porphyrin and the nature of the reaction medium. Consequently, caution should be exercised in generalizing mechanistic conclusions drawn by photophysical and photochemical studies performed under a given set of experimental conditions. This fact is particularly important in porphyrin-sensitized photoreactions occurring in cells or tissues, where porphyrins can be bound at different sites or exist in a large variety of microenvironments (e.g., see Cozzani et al., 1981). Thus, coproporphyrin, which is fairly efficient photosensitizer toward various cell constituents, appears to be ineffective in inducing the photohemolysis of erythrocytes (Haining _et al_., 1969). Analogously, deuteroporphyrin displays a remarkable activity in sensitizing the photokilling of leukemia cells (Kessel, 1977), although it is a poor photosensitizer for several biomolecules (see below).

The experimental conditions (polarity of the medium, ionic strength, temperature etc.) also influence the overall efficiency of the photosensitized processes, since they may enhance or depress the decay rate of ^3P, $^1O_2$ and other transients, as well as the degree of physical quenching of such species by ground state porphyrins and/or substrates.

The definition of the mechanism(s) and efficiency of porphyrin-sensitized photoprocesses often represents a complex problem. We are performing a systematic investigation of the photooxidation of suitable substrates sensitized by porphyrins of selected chemical structures either in homogeneous solutions of known polarity or in organized media which mimic specific aspects of _in vivo_ situations. The results hitherto obtained are summarized in this paper.

Table 1. Selected examples of non-$^1O_2$-involving photoprocesses promoted by photo-excited free base porphyrins

REACTION SYSTEM	OBS. REACTIVE INTERMEDIATES	OBSERVATIONS	REFERENCES
Protoporphyrin IX, 1 mM, DMPO, 50 mM, in phosphate buffer at pH 7.0	O_2^-, OH•	OH• formed via interaction of superoxide with Fe^{+++} ions.	Buettner and Oberley, 1979
Hematoporphyrin IX, several metalloporphyrins + DMPO in acetonitrile, pyridine, dimethylsulfoxide solutions	O_2^-, $^1O_2$	Quantum yield for $^1O_2$ and O_2^- generation dependent on solvent and structure of porphyrin	Cox et al., 1979
Hematoporphyrin IX, 0.5 μM, + Trp, 1.96 mM, in aqueous solutions at pH 10 or in aqueous dispersions of SDS, CTAB, Triton X-100	$^1O_2$, e^-	Reactivity of e^- and $^1O_2$ affected by micellar outer charge	Rossi et al., 1981
Hematoporphyrin derivative + Chinese hamster cells or Micrococcus radiodurans in aqueous medium at pH 7.0	OH•	Production of OH• is indicated from formation of thymine glycols in DNA	Hariharan et al., 1980

EXPERIMENTAL APPROACHES

The following porphyrin-sensitized photoprocesses are used to elucidate the influence of various factors on the efficiency and mechanism of porphyrin-photosensitization:

i) For purely type I photoprocesses: the photoreduction of the N-oxide CTMP to the corresponding amine in deaerated solutions; the progress of the reaction is followed by EPR spectroscopy. Under our experimental conditions (pH 10, 54 μM CTMP), the photoprocess specifically proceeds by e^- transfer from [^3P] (Cannistraro et al., 1981).

ii) For purely type II ($^1O_2$) photoprocesses: the photooxidation of 3.9 mM TMPH, a tertiary amine, measured by EPR; the photooxidation (pH 6-10) of 0.1 mM histidine, measured by spectrophotometric analysis after treatment of irradiated systems with diazo -1H-tetrazole (Horinishi et al., 1964) and 1 mM methionine, measured with ninhydrin (Cauzzo et al., 1977). At the substrate concentrations used by us, type I mechanisms cannot compete with energy transfer from [^3P] to oxygen (Foote, 1976).

iii) For mixed type I/type II photoprocesses: the photooxidation of L-tryptophan (0.1 - 2 mM), followed by the spectrophotofluorimetric procedure described by Genov and Jori (1973). Foote (1976) calculated that, in the presence of 0.25 mM Trp and in air-equilibrated aqueous solutions, the triplet state of methylene blue has equal probability of being deactivated by type I and II mechanisms..

The presence of the potential intermediate species and the relative importance of the single reaction pathways is estimated by the addition of selective quenchers or enhancers to the irradiated system: in particular, we use NaN_3 and D_2O as a quencher and enhancer, respectively, of $^1O_2$, potassium ferricyanide and cysteine as scavengers of e^-, superoxide dismutase as a scavenger of $O_2^{\cdot -}$, and hydroquinone as a radical species scavenger.

The irradiations are performed by using either a high pressure Xenon lamp (Osram, 500 W) or four tungsten bulbs (Phillips, 250 W) as the light source. Wavelengths below 400 nm are eliminated by placing chemical or interference filters in the front of the irradiated solutions. In kinetic experiments, there is total absorption of the incident light by the porphyrin sensitizer.

As usual for dye-sensitized photooxidations, photoprocesses promoted by photoexcited porphyrins follow first-order kinetics with respect to substrate concentration (Jori et al., 1969). The first order rate constant, deduced from the slope of a semilogarithmic plot: ln [substrate] vs. time, is used to evaluate the efficiency of the photoprocesses.

STUDIES IN HOMOGENEOUS SOLUTIONS

1. Photooxidations sensitized by porphyrins of different chemical structure.

Three structural features are being examined: i) the presence of metal ions coordinated with the pyrrole nitrogens; ii) the nature of the side chains departing from the 1-8 positions of the tetrapyrrolic macrocycle; iii) the presence of substituents in the meso positions.

As shown in Table 2, the metallo-derivatives of HP are remarkably less efficient photosensitizers than free base porphyrin. Flash photolysis and photooxidation kinetic studies at different O_2 pressures indicate that, for Cu^{2+}- and Fe^{3+}-HP, the drop of the photoactivity reflects a drastic shortening of the triplet lifetime (Cauzzo et al., 1977). The effect is probably related to the paramagnetic properties of the bound metal ions, causing an enhancement of the spin-orbit coupling, since the presence of two CN^- ions in the fifth and sixth coordination positions of Fe^{3+}, which renders the system diamagnetic, endows the porphyrin with photosensitizing efficiency, at least in $^1O_2$-involving photoprocesses (see table 2; Cannistraro et al., 1978a). The lifetime of triplet HP is shortened to a lesser extent by the insertion of diamagnetic Zn^{2+} and Mg^{2+} ions; in this case, the photosensitizing efficiency is also lowered by the increased internal conversion from the first excited singlet state to the ground state: actually, Zn^{2+} and Mg^{2+} HP display both fluorescence and intersystem crossing quantum yields lower than those of HP (Cauzzo et al., 1977).

All metallo-derivatives of HP have a relatively greater tendency to initiate e^- transfer reactions than to generate $^1O_2$. According to Ballard and Mauzerall (1980), the e^- donating ability of metalloporphyrins is related to the low probability of back reaction between the initially formed cation and anion radicals. Thus, the fact that the Mg^{2+} and Zn^{2+} derivatives are about 50% as efficient as HP in promoting Trp photooxidation, but less than 10% as efficient in promoting the photooxidation of typical $^1O_2$ -acceptors, strongly suggests that Mg^{2+} and Zn^{2+}-HP sensitize Trp photooxidation mainly via e^--transfer.

As regards the influence of substituents in the angular and meso positions, the data summarized in Table 3 point out that, out of several dicarboxylic porphyrins, HP is most ef-

Table 2. Relative efficiency of various metalloporphyrins as photosensitizers

PORPHYRIN	TYPE I PHOTOPROCESS substrate: CTMP	TYPE II PHOTOPROCESS substrates		TYPE I/II PHOTOPROCESS substrate: TRP
		MET	TMPH	
HP	1.00	1.00	1.00	1.00
Mg^{2+}-HP	0.46	0.18	0.16	0.08
Zn^{2+}-HP	0.40	0.03	0.04	0.46
Cu^{2+}-HP	0.15	0.01	0.00	---
Fe^{3+}-HP	0.00	0.00	0.00	---
Fe^{3+}-PP†	---	0.00	---	0.00
Di-cyan-Fe^{3+}-PP	---	0.16	0.16	0.27

†Protoporphyrin

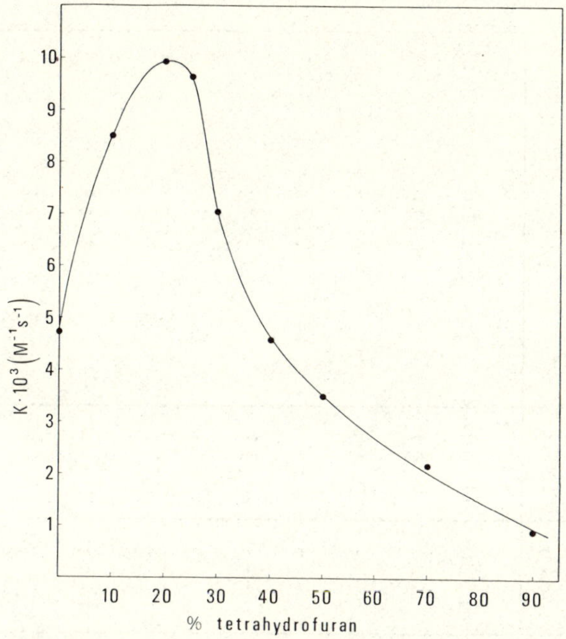

Figure 1. First-order rate constant for the photooxidation of 0.1 mM L-tryptophan sensitized by 0.083 mM HP in solvent mixtures containing specified percentages of tetrahydrofuran and 0.1M phosphate buffer pH 7.4.

ficient in promoting both $^1O_2$ and e^- transfer photoprocesses. Increasing the number of side chain carboxylate functions or introducing hydrophilic meso substituents (both positively and negatively charged) enhances the efficiency of $^1O_2$ photoprocesses. Apparently, the efficiency of Trp photooxidation also increases upon increasing hydrophilicity of the angular or meso substituents, which may indicate some role of $^1O_2$ in the modification of this target. Inhibition studies with azide and ferricyanide are in progress in order to assess the validity of this hypothesis.

Since most free base porphyrins display essentially similar intersystem crossing quantum yields and triplet lifetimes (Bonnett et al., 1980), the observed differences in the efficiency of pure $^1O_2$ photoprocesses promoted by different porphyrins (Table 3) may reflect different degrees of $^1O_2$ physical quenching by ground state dye, as it has been recently estimated for chlorophyll and other pigments (Tanielan and Golder, 1981). In some cases (e.g., protoporphyrin), $^1O_2$ can also chemically react with the porphyrin, which is converted to products devoid of photosensitizing

Table 3. Relative efficiencies of free base porphyrins as photosensitizers

PORPHYRIN	TYPE I PHOTOPROCESS substrate: CTMP	TYPE II PHOTOPROCESS substrates			TYPE I/II PHOTOPROCESS substrate: TRP
		MET	HIS	TMPH	
Hematoporphyrin IX	1.00	1.00	1.00	1.00	1.00
Protoporphyrin IX	0.11	0.11	---	0.07	0.10
Deuteroporphyrin IX	0.05	0.19	---	0.16	---
Mesoporphyrin IX	0.09	---	---	---	0.21
Protoporphyrin IX dimethylester	---	0.22	---	0.17	---
Mesoporphyrin IX dimethylester	---	0.28	---	0.23	---
Coproporphyrin III	---	---	0.71	---	1.23
Uroporphyrin III	---	---	3.27	---	1.41
TPPS	---	---	3.56	---	2.02
TCPP	---	---	3.10	---	1.53
PAP	---	---	2.10	---	---

activity (Hopf and Whitten, 1975). The various porphyrins may also have different triplet energy levels, which in turn affects the efficiency of electronic energy transfer to $^3O_2$.

2. Porphyrin-sensitized photooxidation of tryptophan in media of different polarity.

The HP-sensitized photooxidation of Trp has been studied in solvent mixtures containing different molar fractions of water and water-miscible organic solvents, whose dielectric constant ε is either lower or higher than that of water, in order to assess the importance of medium polarity in controlling HP-sensitized photoprocesses.

The following organic solvents were used: formamide (ε=109 at 20°C), N,N-dimethylformamide (38.05), methanol (32.63), ethanol (24.31), and tetrahydrofuran (2.95). In media less polar than water, plotting the first-order rate constant for Trp photooxidation vs. the percentage of organic component yields bell-shaped plots (Fig. 1). Spectroscopic studies point out that the ascending portion of the plot is coincident with the prevalence of aggregated HP species, whereas the descending portion corresponds to the presence of HP as a monomer (Jori et al., 1980). Our results are summarized in Table 4 and have been recently confirmed by Kessel and Rossi (1981).

Several results (Jori et al., 1980; Tomio et al., 1980) suggest that aggregated HP exerts its photosensitizing action mainly via $^1O_2$. Thus, the rate of the HP-sensitized photooxidation of His and Met (a pure $^1O_2$-photoprocess) in water-methanol solvent mixtures steadily increases upon decreasing solvent polarity, as one should expect owing to the higher solubility of $^3O_2$ and the longer lifetime of $^1O_2$ in methanol than in water (Merkel et al., 1972). Moreover, N_3^- quenching of Trp photooxidation becomes progressively more efficient in the 0-30% methanol range, and becomes less important at higher methanol concentrations. The change of the slope of the plot above a given organic solvent concentration would indicate that monomeric HP sensitizes Trp photooxidation mainly via e^- transfer. Actually, we observe a decreasing efficiency of the overall photoprocess upon decreasing medium polarity; this can be ascribed to an enhanced probability of back electron transfer from Trp radical anion to the HP radical cation. In water-formamide mixtures (Table 4) the efficiency of the photoprocess promoted by monomeric HP remains essentially constant because of stabilization of the intermediate radical ions in media of high polarity.

The above outlined experiments have been extended to other porphyrins and metalloporphyrins in water-methanol and water-formamide mixtures (Reddi et al., 1981), as shown in

Table 4. Effect of solvent composition on hematoporphyrin-sensitized photooxidation of tryptophan in water-organic solvent mixtures

ORGANIC SOLVENT (ε at 25°C)	PHYSICAL STATE OF HP	TYPE OF PLOT OBTAINED*	R_{MAX}‡
N,N-dimethylformamide (38.05)	60%+	Bell-shaped	60%
Methanol (32.63)	30%	Bell-shaped	30%
Ethanol (24.31)	25%	Bell-shaped	20%
Tetrahydrofuran (2.95)	20%	Bell-shaped	20%
Formamide (109)	30%	Sigmoidal	40-90%

*Plot of first-order rate constant for Trp photooxidation vs. percentage of organic phase in the solvent mixture.

‡%Organic Solvent for maximal reaction rate.

+Porphyrin aggregated if % organic solvent <this value.

Tables 5 and 6. Clearly, in the region of low organic solvent concentration, there is a very good correspondence between the interval of preferential existence of aggregated porphyrins and the ascending portion of bell-shaped or sigmoidal plots. The rate of photosensitization by uroporphyrin, which is monomeric in aqueous solutions, decreases steadily upon addition of methanol to the solvent system. Hence, the conclusion that aggregated porphyrins exert their photosensitizing action mainly via $^1O_2$, while monomeric porphyrins preferentially promote e^- transfer processes, appears to be of general validity.

STUDIES IN ORGANIZED MEDIA

1. Sensitized photooxidations by protein-bound porphyrins.

Serum proteins, including albumin and hemopexin, are the natural carriers of porphyrins to liver and other tis-

Table 5. Plots obtained by representing the first-order rate constant for the photooxidation of tryptophan sensitized by various porphyrins in water-methanol mixtures as a function of the solvent composition.

PORPHYRIN	PHYSICAL STATE OF THE PORPHYRIN	TYPE OF PLOT OBTAINED	R_{MAX}‡
HP	30%*	Bell-shaped	30%
Mg^{2+}-HP	40%	Bell-shaped	40%
Zn^{2+}-HP	60%	Bell-shaped	40%
Mesoporphyrin	50%	Bell-shaped	50%
Coproporphyrin	30%	Bell-shaped	25%
Uroporphyrin	†	¶	0
TCPP	30%	Bell-shaped	30%
TPPS	20%	Bell-shaped	20%

‡% organic solvent for maximal reaction rate.

*Porphyrin aggregated at % formamide < this value. exceed this value.

†Monomeric at 0-100% methanol.

¶Steady decrease with increasing % methanol, inflection at 50%.

sues (Koskelo and Muller-Eberhard, 1977), including neoplastic tissues (Jori et al., 1979). Under physiological conditions, HP forms a 1:1 complex HSA; with stability constant = 3 x 10^6 M^{-1} (Reddi et al., 1981b). Since specific membrane proteins may associate with porphyrins, (Girotti, 1979; Kohn and Kessel, 1979), we examined effects of protein binding on porphyrin photosensitization efficiency.

We studied the kinetics of the photooxidative degradation of the single Trp residue of HSA as promoted by bound HP and its Mg^{2+}- and Zn^{2+}-derivatives. The intramolecular distance between Trp and the bound porphyrins was found to be 1.8-2 nm by electronic energy transfer studies (Reddi et al., 1981b). For comparison, the model compound NATA was

Table 6. Plots obtained by representing the first-order rate constant for the photooxidation of tryptophan sensitized by various porphyrins in water-formamide mixtures as a function of the solvent composition

PORPHYRIN	PHYSICAL STATE OF THE PORPHYRIN	TYPE OF PLOT OBTAINED	R_{MAX}‡
HP	30%*	Sigmoidal	40-90%
Mg^{2+}-HP	40%	Quasi-sigmoidal	¶
Zn^{2+}-HP	30%	Sigmoidal	30-90%
Mesoporphyrin	60%	Sigmoidal	60-90%
Coproporphyrin	30%	Bell-shaped	30%
Uroporphyrin	†	Sigmoidal	20-90%
TCPP	50%	Sigmoidal	50-90%
TPPS	40%	Sigmoidal <50, then decreasing	---

‡% Organic solvent for maximal reaction rate

*Porphyrin aggregated at % formamide < this value.

†Porphyrin monomeric in 0-100% formamide.

¶Inflection at 40%, maximal rate at 80%.

irradiated in the presence of free porphyrins under identical experimental conditions. As shown in Table 7, protein-bound porphyrins are 2-4.5-fold more efficient than unbound porphyrins in inducing Trp photooxidation. On the contrary, methylene blue, which does not bind with HSA, sensitizes the photooxidation of free Trp in NATA at a rate ($10.14 \times 10^{-3} s^{-1}$) by far larger than that typical of an albumin Trp residue ($4.35 \times 10^{-3} s^{-1}$).

The enhancement of the photosensitizing efficiency is generally observed for dyes bound to macromolecular structures (Jori and Spikes, 1981). This effect has been ascribed to the reduced probability of collisional deactiva-

Table 7. Rate constants for the photooxidation of the single Trp residue of HSA by HSA-bound porphyrins and of the Trp moiety of NATA by free porphyrins.

PORPHYRIN	RATE CONSTANT (s^{-1})		RATIO bound/free
	bound porphyrin	free porphyrin	
HP	6.54×10^{-3}	2.97×10^{-3}	2.21
Mg^{2+}-HP	3.43×10^{-3}	0.77×10^{-3}	4.47
Zn^{2+}-HP	4.79×10^{-3}	2.43×10^{-3}	1.97

tion of the electronically excited states of bound sensitizers (e.g., dyes located in hydrophobic regions of protein molecules often show phosphorescence emission even at room temperature). In particular, binding with HSA significantly increases the fluorescence quantum yield of HP. Moreover, the red shift of the absorption spectrum typical of protein-bound porphyrins (Koskelo and Muller-Eberhard, 1977) implies a smaller singlet-triplet splitting, which in turn increases the efficiency of intersystem crossing to the lowest triplet state. At present, we are extending the above outlined investigations to ground state complexes between HP and selected cell constituents. In any case, the relevance of these findings for studies of porphyrin photosensitization in vivo is emphasized by literature reports stating a higher efficiency of cell-bound porphyrins in inducing photodamage (Kohn and Kessel, 1979; Sandberg and Romslo, 1981).

2. Porphyrin-sensitized photooxidations in aqueous micellar systems.

One feature of photosensitized processes occurring in cells is often represented by the lack of physical interaction between substrate and sensitizer. The reactivity and diffusability of the photogenerated intermediates may be affected by abrupt phase changes and variations in medium polarity because of the highly heterogeneous and compartmentalized structure of the system. Such aspects can be mimicked by aqueous micellar dispersions owing to the presence of both hydrophobic and hydrophilic regions in these amphipathic surfactant assemblies. In particular, HP is readily incorporated in a monomeric state into anionic (SDS), cationic (CTAB), and neutral (Triton X-100) micelles, whereas Trp and its decarboxylated analogue tryptamine partition either in the bulk aqueous phase or in the micellar core, depending on the outer charge of the micelle (Sconfienza et al., 1980; Rossi et al., 1981). Therefore, by irradiation of these systems, it is possible to explore different types of interactions between the porphyrin sensitizer and the substrate.

Actually, micellized HP does sensitize the photooxidation of Trp and tryptamine whether they are in the aqueous medium or lie in a different micelle. Inhibition studies indicate that both e^- and $^1O_2$ are involved in the photoprocess, although the relative weight of either reaction pathway is strongly dependent on the experimental conditions. Our findings receive some support from the literature. Almgren and Thomas (1980) demonstrated that, in aqueous dispersions of CTAB, electrons can visit a large number of micelles before reacting or being deactivated; analogously, $^1O_2$ can diffuse out of the original micelle through the aqueous phase and penetrate another substrate-containing micelle (Lindig and Rodgers, 1981).

The data obtained in the HP-sensitized photooxidation of Trp and tryptamine are summarized in Tables 8 and 9, respectively. On the whole, our results suggest the following conclusions:

i) The reactivity of $^1O_2$ toward substrates located in the aqueous milieu is only slightly affected by empty micelles, as shown by the similar degree of D_2O acceleration and N_3^- inhibition of Trp photooxidation in homogeneous solutions and in aqueous dispersions of SDS and Triton micelles.

ii) The efficiency of $^1O_2$ photoprocesses is increased by the association of Trp with CTAB micelles. Spectroscopic studies indicate that this effect is due to a favourable orientation of the indole side chain of Trp toward the attacking species. Similar considerations obtain for the higher importance of the $^1O_2$ involving mechanism in the photooxidation of CTAB-associated tryptamine (Rossi et al., 1981).

iii) The incorporation of the substrate into anionic micelles drastically prevents its reaction with e^- (e.g., ferricyanide has practically no effect on the HP-sensitized photooxidation of tryptamine in SDS micelles) probably as a consequence of electrostatic repulsion. If the substrate is in the aqueous phase, the anionic micelle favours the e^- transfer mechanism by preventing back transfer (Wallace and Thomas, 1973).

iv) The e^- readily attacks substrates located in cationic or neutral micelles. Once again, the efficiency of

Table 8. Effect of various quenchers and enhancers on the HP-sensitized photooxidation of tryptophan in aqueous dispersions of neutral or ionic micelles at pH 10.

REACTION SYSTEM	RATE CONSTANT x 10^3 (s^{-1})	ACCELERATING EFFECT OF D_2O	$[NaN_3]$ FOR 50% INHIBITION	% QUENCHING BY FERRICYANIDE 0.5 mM
$(HP)_{aq}$ - $(Trp)_{aq}$	5.21	2.23-fold	7.50 mM	72
$(HP)_{SDS}$ - $(Trp)_{aq}$	6.23	1.66-fold	8.00 mM	32
$(HP)_{Tri}$ - $(Trp)_{aq}$	8.37	2.06-fold	3.31 mM	47
$(HP)_{CTAB}$ - $(Trp)_{CTAB}$	2.75	4.00-fold	0.40 mM	43

Experimental conditions: Trp, 1.96 mM; HP, 0.51 μM; Surfactants, 20 mg/ml. All experiments were run in the presence of 0.1 M Tris (pH 10).

Table 9. Effect of various quenchers and enhancers on the hematoporphyrin-sensitized photooxidation of tryptamine in aqueous dispersions of neutral or ionic micelles at pH 10.

REACTION SYSTEM	RATE CONSTANT ($x10^3$ s^{-1})	ACCELERATING EFFECT OF D_2O	$[NaN_3]$ FOR 50% INHIBITION	% QUENCHING BY FERRICYANIDE 0.5mM
$(HP)_{aq}$ - $(TAM)_{aq}$	4.53	6.20-fold	0.40 mM	46
$(HP)_{SDS}$ - $(TAM)_{SDS}$	5.04	1.61-fold	2.00 mM	2
$(HP)_{Tri}$ - $(TAM)_{Tri}$	5.98	2.72-fold	0.65 mM	31
$(HP)_{CTAB}$ - $(TAM)_{CTAB}$	2.61	5..56-fold	0.40 mM	18

Experimental conditions: TAM, 0.5 mM; HP, 0.51 μM; surfactants, 20 mg/ml. All experiments were run in 0.1 M Tris (pH 10).

this reaction pathway is controlled by the orientation of the substrate in the micelle: e.g., the -amino group of Tryptamine, which is the main site of attack by e^-, lies closer to micellar surface in Triton than in CTAB micelles (Rossi et al., 1981)

ACKNOWLEDGEMENTS

The investigations described in this paper were supported in part by C.N.R. under the Progretto Finalizzato "Controllo della Crescita Tumorale Maligna" and in part by the Provincia di Padova. One of us (Elena Reddi) is the recipient of a fellowship given by the Provincia di Padova.

REFERENCES

1. Almgren, M. and J.K. Thomas (1980) Interfacial electron transfer involving radical ions of carotene and diphenylhexatriene in micelles and vesicles. Photochem. Photobiol. 31, 329-335.

2. Ballard, S.G. and D.C. Mauzerall (1980) Photochemical ionogenesis in solutions of zinc octaethylporphyrin. J.Chem. Phys. 72, 933-947.

3. Bonnett, R., A.A. Charalambides and E.J. Land (1980) Triplet states of porphyrin esters. J.C.S. Faraday I, 76, 852-859.

4. Buettner, G.R. and L.W. Oberley (1979) Superoxide formation by protoporphyrin as seen by spin trapping. FEBS Lett. 98, 18-20.

5. Cannistraro, S., G. Jori and A. Van de Vorst (1978a) Photosensitization of amino acids by di-cyan-hemin: kinetic and EPR studies. Photochem. Photobiol. 27, 257-260.

6. Cannistraro, S., A. Van de Vorst and G. Jori (1981) Quantum yield of electron transfer and of singlet oxygen production by porphyrins: an EPR study. Photobiochem. Photobiophys., in press.

7. Cauzzo, G., G. Gennari, G. Jori and J.D. Spikes (1977) The effect of chemical structure on the photosensitizing efficiency of porphyrins. Photochem. Photobiol. 25, 389-395.

8. Cox, G.S., D.G. Whitten and C. Giannotti (1979) Interaction of porphyrin and metalloporphyrin excited states with molecular oxygen. Energy-transfer versus electron-

transfer quenching mechanisms in photooxidations. Chem. Phys. Lett. 67, 511-515.

9. Cozzani, I., G. Jori, E. Reddi, A. Fortunato, L. Tomio and P.L. Zorat (1981). Distribution of endogenous and injected porphyrins at the subcellular level in rat hepatocytes and in ascites hepatoma. Chem.-Biol. Interactions, in press.

10. Dougherty, T.J., C.J. Gomer and K.R. Weishaupt (1976) Energetics and efficiency of photoinactivation of murine tumor cells containing hematoporphyrin. Cancer Res. 36, 2330-2333.

11. Foote, C.S. (1976) "Photosensitized oxidation and singlet oxygen: consequences in biological systems." In: Free Radicals in Biology (ed., W.A. Pryor), Vol. II, Academic Press, New York, pp.85-133.

12. Genov, N. and G. Jori (1973) Conformational studies on the alkaline protease from Bacillus mesentericus. Int. J. Peptide Protein Res. 5, 127-133.

13. Girotti, A.W. (1979) Protoporphyrin-sensitized photodamage in isolated membranes of human erythrocytes. Biochem. 18, 4403-4409.

14. Haining, R.G., E. Hulset and R.F. Labbe (1969) Photohemolysis. The comparative behaviour of erythrocytes from patients with different types of porphyria. Proc. Soc. Expl. Biol. Med. 132, 625-628.

15. Hariharan, P.V., J. Courtney and S. Eleczko (1980) Production of hydroxyl radicals in cell systems exposed to hematoporphyrin and red light. Int. J. Radiat. Biol. 37, 691-694.

16. Hopf, F.R. and D.G. Whitten (1975) "Photochemistry of porphyrins and metalloporphyrins" in: Porphyrins and Metalloporphyrins (ed., K.M. Smith), Elsevier Publishing Co., Amsterdam, pp. 667-700.

17. Horinishi, H., Y. Hachimori, K. Kurihara and K. Shibata (1964) States of amino acid residues in proteins. III. Histidine residues in insulin, lysozyme albumin and proteinases as determined with a new reagent of diazo-1H-tetrazole. Biochim. Biophys. Acta 86, 477-489.

18. Ito, T. (1978) Cellular and subcellular mechanisms of photodynamic action: the singlet oxygen hypothesis as a driving force in recent research. Photochem. Photobiol. 28, 493-508.

19. Jori,G. , G.Galiazzo and E. Scoffone (1969) Photodynamic action of porphyrins on amino acids and proteins. I. Selective photooxidation of methionine in aqueous solution. Biochem. 8, 2969-2875.

20. Jori,G., G.B. Pizzi, E. Reddi, L. Tomio, B.Salvato, P.L.Zorat and F.Calzavara (1979) Time-course of hematoporphyrin distribution in selected tissues and in ascites hepatoma of normal and tumour-bearing rats. Tumori 65, 43-52.

21. Jori,G., E. Reddi, E. Rossi, I. Cozzani, L. Tomio, P.L. Zorat, G.B. Pizzi and F. Calzavara (1980) Porphyrin-sensitized photoreactions and their use in cancer phototherapy. Med. Biol. Environ. 8, 139-154.

22. Jori,G. and J.D. Spikes (1981) "Photosensitized oxidations in complex biological structures", In: Oxygen and Oxy-radicals in Chemistry and Biology, (ed., M.A.J. Rodgers and E.L. Powers), Academic Press, New York, pp 441-453.

23. Kessel,D. (1977) Effects of photoactivated porphyrins at the surface of leukemia L1210 cells. Biochem. 16:3443-3449.

24. Kessel,D. and E. Rossi (1982) Determinants of porphyrin-induced photooxidation characterized by fluorescence and absorption spectra. Photochem. Photobiol., 35, 37-41.

25. Kohn, K. and D. Kessel (1979) On the mode of cytotoxic action of photoactivated porphyrins. Biochem. Pharmacol. 28, 2465-2470.

26. Koskelo,P. and U. Muller-Eberhard (1977) Interaction of porphyrins with proteins. Seminars Hematol. 14, 221-226.

27. Lindig,B. and M.A.J. .Rodgers (1981) Rate parameters for quenching of singlet oxygen by water-soluble and lipid-soluble substrates in aqueous and micellar systems. Photochem. Photobiol. 33, 627-634.

28. Maillard,P., P. Krausz, C. Giannotti and S. Gaspard (1980) Photoinduced activation of molecular oxygen by various porphyrins, bis-porphyrins, phthalocyanines, pyridinoporphyrazines and their metal derivatives. J. Organomet. Chem. 197, 285-290.

29. Merkel,P.B., R. Nilsson and D.R. Kearns (1972) Remarkable solvent effect on the lifetime of singlet oxygen. J. Am. Chem. Soc. 94, 1029-1030.

30. Reddi,E., E. Rossi and G. Jori (1981a) Factors control-

ling the efficiency of porphyrins as photosensitizers of biological systems to damage by visible light. Med. Biol. Environ. 9, 337-351.

31. Reddi,E., F. Ricchelli and G. Jori (1981b) Interaction of human serum albumin with hematoporphyrin and its Zn^{2+} and Mg^{2+} derivatives. Int. J. Peptide Prot. Res., in press.

32. Rossi,E., A. van de Vorst and G. Jori (1981) Competition between the singlet oxygen and electron transfer mechanisms in the porphyrin-sensitized photooxidation of L-tryptophan and tryptamine in aqueous micellar dispersions. Photochem. Photobiol. 34, 447-454.

33. Sandberg,S. and I. Romslo (1981) Porphyrin-induced photodamage at the cellular and subcellular level as related to the solubility of the porphyrin. Clin. Chim. Acta 109, 193-201.

34. Spikes,J.D. (1975) Porphyrins and related compounds as photodynamic sensitizers. Ann. N. Y. Acad. Sci. 244, 496-508.

35. Tanielian,C. and L. Golder (1981) Quenching of singlet oxygen by sensitizer in dye-sensitized photooxygenation. Photochem. Photobiol. 34, 411-414.

36. Wallace,S.C. and J.K. Thomas (1973) Reactions in micellar systems. Radiat. Res 54, 49-62.

PHOTODYNAMIC ACTION OF PROTOPORPHYRIN ON RESEALED ERYTHROCYTE MEMBRANES: MECHANISMS OF RELEASE OF TRAPPED MARKERS

Albert W. Girotti and Mark R. Deziel

Department of Biochemistry
The Medical College of Wisconsin
Milwaukee, Wisconsin 53226 U.S.A.

INTRODUCTION

When administered to tumor-bearing animals, certain porphyrin photosensitizers, most notably "hematoporphyrin derivative (HPD), tend to become localized in the tumor area. This tissue is then susceptible to selective destruction when the sensitizer is photoactivated by visible light (Dougherty et al., 1975). The mechanisms of porphyrin targeting and cytotoxicity are poorly understood and are currently the subjects of intensive investigation. Recent studies on neoplastic cells in culture have suggested that loss of viability may be closely linked with structural and functional derangements in the plasma membrane (Kessel, 1977; Kohn and Kessel, 1979). However, the exact molecular factors involved have not been elucidated. We have been investigating porphyrin-sensitized photodamage in the human red blood cell (RBC) membrane as a simple model for understanding porphyrin phototoxicity in general. Some of this work may provide new insights into the tumoricidal aspects of this phenomenon. In the present report we describe the damaging effects of photoactivated protoporphyrin (PP) on resealed RBC ghosts and characterize the molecular lesions which are responsible for increased permeation of small solute markers of membrane integrity.

MATERIALS AND METHODS

Materials. Monosodium D-glucose-6-phosphate (G6P), $NADP^+$, glucose-6-phosphate dehydrogenase (G6PDH, Type VII from yeast), p-chloromercuribenzenesulfonic acid (PCMBS), and 2,6-di-t-butyl-4-methylphenol (butylated hydroxytoluene, BHT) were purchased from Sigma Chemical Co., St. Louis, MO. Protoporphyrin IX dimethyl ester was a product of Schwarz/ Mann, Orangeburg, NY. Free protoporphyrin (PP) was prepared by acid hydrolysis of the ester (Mauk and Girotti, 1973). N,N,N',N'-tetramethyl-p-phenylenediamine (TMPD) and 1,6-diphenylhexatriene (DPH) were products of Eastman Organic Chemicals (Rochester, NY) and Aldrich Chemical Co. (Milwaukee, WI), respectively. Thin layer chromatography (TLC) plates (Silica gel-G, 0.25 mm thickness) were from Supelco Inc., Bellefonte, PA. The remaining chemicals were reagent grade and all solutions were prepared with glass-distilled water.

Freshly drawn human blood was provided by the Blood Center of Southeastern Wisconsin, Milwaukee, WI. The procedure for preparing rightside-out resealed RBC ghosts (essentially hemoglobin-free) was described previously (Deziel and Girotti, 1980a,b). Trapped Na^+ and G6P were used as permeability markers. The final bathing medium was usually PBS [145 mM NaCl + 10 mM sodium phosphate (pH 7.4)]. For experiments in which Na^+ efflux would be measured, the ghosts were washed with and resuspended in PBC [145 mM choline chloride + 10 mM potassium phosphate (pH 7.4)].

Illumination. Stock solutions of PP (100 μM in 2 mM KOH) were prepared immediately before use. The porphyrin was slowly mixed with the membrane suspension [1.0 mg of protein (approx. 2 x 10^9 ghosts)/ml in PBS or PBC] to give a desired final concentration (typically 2 μM). After 15 min of dark incubation, the stirred suspensions, 5-6 ml each in paired, 10°C thermostatted beakers (45 mm diameter) were irradiated, using R-40 mercury arc lamps (Duro-Test Corp., North Bergen, NJ) and corning CS 5-57 filters (blue transmitting). Suspensions were in contact with air at all times. Incident light intensity was measured with a YSI Radiometer (YSI Instruments, Yellow Springs, OH).

Permeability measurements. A schematic depicting marker escape and its quantitation is shown in Fig. 1. Emitted Na^+ was determined with a specific ion electrode (Orion 96-11) attached to a Radiometer pH Meter 26 operating in the millivolt mode. G6P (the dianion at pH 7.4) was monitored spectrophotometrically at 340 nm by coupling its G6PDH-catalyzed oxidation to 6-phosphogluconolactone (6PGL) with the reduction of $NADP^+$ (cf. Deziel and Girotti, 1980a,b for details). In principle, external markers can be determined using whole suspension samples. However, turbidity of

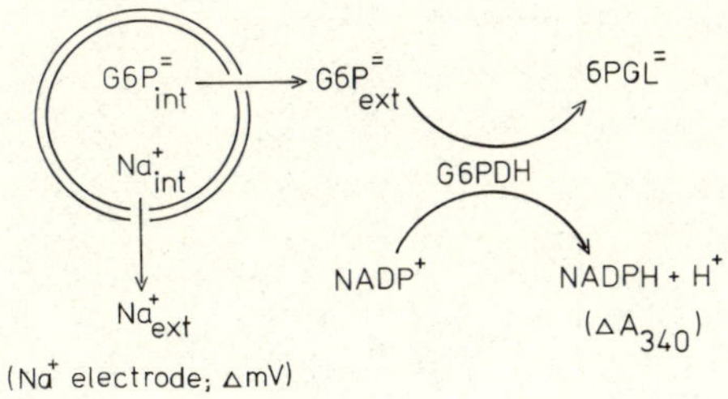

Figure 1. Measurement of Marker Efflux.

the ghost suspensions necessitated centrifugation prior to spectrophotometric measurements. Total marker concentrations were obtained by lysing ghosts with Triton X-100 (0.1%).

Measurements of lipid peroxidation. Lipid peroxidation in irradiated ghosts was measured colorimetrically by a recent modification of the classic thiobarbituric acid assay (Asakawa and Matsushita, 1980). This assay detects carbonyl breakdown products of hydroperoxide intermediates. An aliquot of membrane sample (0.15 mg of protein) was added to a 2 ml assay mixture containing 0.3% 2-thiobarbituric acid (TBA), 0.2% sodium dodecyl sulfate, 50 μM $FeCl_3$, and 0.5 mM BHT in 0.1 M glycine-HCl, pH 3.6. The samples were placed in a boiling water bath for 10 min, cooled to room temperature and the absorbance at 532 nm was then measured. Absorbance readings were corrected for sample blanks (membranes solubilized in dodecyl sulfate) and reagent blanks (lacking membranes). Lipid peroxidation was also assessed by direct detection of individual hydroperoxides on thin layer chromatograms. Membrane samples (0.5 ml) were mixed with 0.3 ml $CHCl_3$-CH_3OH (2:1, v/v), vortexed, and centrifuged for 5 min at 13,000 x g. Aliquots of 50 μl from the lower phase were applied to Silica gel-G plates along with non-oxidized cholesterol and phospholipid standards. The plates were developed with $CHCl_3$-CH_3OH-H_2O (30:10:1.6, v/v/v), air dried for a few minutes, and then sprayed lightly with TMPD [1% (w/v) in 50% aqueous methanol containing 1% glacial acetic acid]. Reaction with hydroperoxides converts TMPD to the intensely purple Wurster dye (Smith and Hill, 1972). After photographic recording of the results, the plates were sprayed with the fluorophore DPH [0.01% (w/v) in petroleum ether] for general detection of all lipid classes (Hyslop and York, 1980). Excitation with long wavelength ultraviolet light allowed visualization of the parent lipid spots. (Oxidation products showed little or no fluorescence). The plates were finally sprayed with ninhydrin

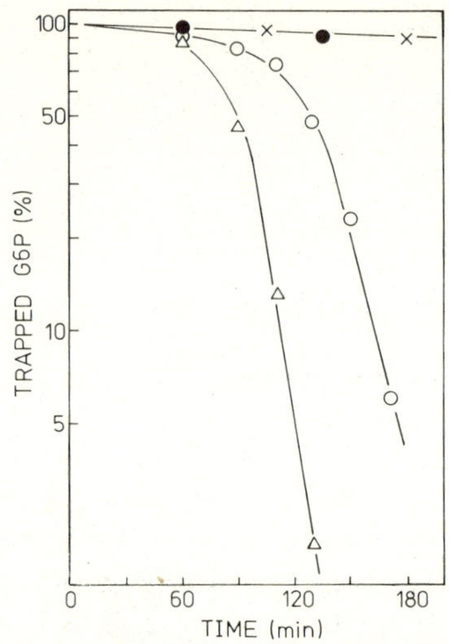

Figure 2. Photosensitized lysis of resealed RBC ghosts; effect of porphyrin dose. External G6P was monitored during continuous irradiation in the presence of 2 μM PP (o) and 5 μM PP (Δ). Light intensity was 25 W/m^2. Efflux in a dark control containing PP (•) and in a light control lacking porphyrin (**x**) is identical with background efflux at 10°C in normal room light with no added sensitizer.

[0.1% (w/v) in ethanol] and gently heated to confirm the identity of the amino-lipids phosphatidylethanolamine and phosphatidylserine.

RESULTS

Irradiation of resealed ghosts in the presence of PP resulted in the eventual total release of trapped Na^+ and G6P. As shown in Fig. 2, efflux of G6P at the given light intensity (25 W/m^2) was dose dependent with respect to sensitizer concentration. At 2 μM PP the marker emerged abruptly (and in apparent first order fashion) after a lag period of approximately 90 min. Increasing the porphyrin concentration to 5 μM shortened the lag and increased the efflux rate. When light was removed at some point during the rapid release phase, efflux returned to its basal rate within about 15 min (not shown). These results demonstrate that the membrane damaging effects of photoactivated PP are manifested throughout the entire period of irradiation.

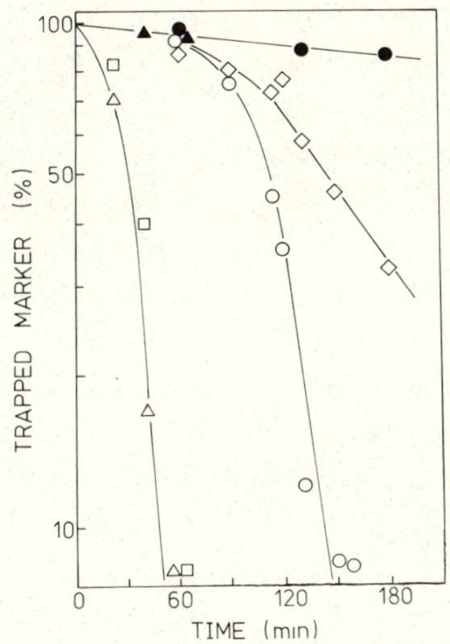

Figure 3. Comparison of Na^+ and G6P release from irradiated ghosts. The ghosts were sensitized with 2 μM PP and irradiated in the absence of BHT (△, Na^+; ○, G6P) and in the presence of 0.5 mM BHT (□, Na^+; ◇, G6P). Light intensity was 25 W/m^2. Also shown is the basal efflux of Na^+ (▲) and G6P (●).

The kinetic profile of Na^+ release from photodamaged ghosts differed in several respects from that of G6P (Fig. 3). Sodium efflux was rapid, continuous (no lag), and essentially complete before G6P began to emerge. To learn whether free radical peroxidation of unsaturated lipids plays any role in marker release, we studied the effects of antioxidants (radical traps) on this process. Prior addition of 0.5 mM butylated hydroxytoluene (BHT) to one of the irradiated ghost suspensions shown in Fig. 3 had only a minimal inhibitory effect on Na^+ release, but diminished the rate of G6P release by about 4-fold. (There was no apparent change in the lag period; however, additional evidence will be needed to verify this.) In a separate experiment, the light intensity was reduced 5-fold to approx. 5W/m^2 to allow collection of more data concerning the apparent small inhibition of Na^+ release observed at 25 W/m^2. At the lower intensity BHT had no effect at all on the cation (Fig. 4). As expected, egress of G6P was again suppressed under these conditions (not shown). These results suggest that the molecular lesions responsible for Na^+ and G6P escape are different. On the basis of this and similar evidence obtained previously with another sensitizer (Deziel and Girot-

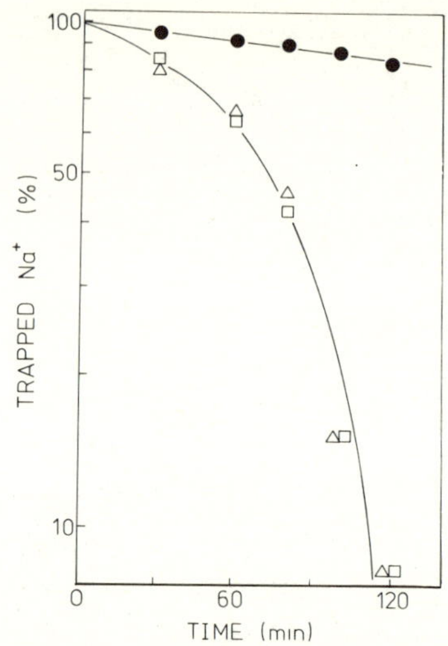

Figure 4. Lack of an antioxidant effect on Na^+ release. Resealed ghosts were sensitized with 2 μM PP and irradiated at a light intensity of 5 W/m². Cation leakage in the absence (△) and presence (□) of BHT (0.5 mM) is plotted against irradiation time. Data from a PP-containing dark control are plotted alongside (•).

Girotti, 1980a,b), we propose that protein photooxidation (non-lipid dependent) is sufficient to emit Na^+. (The small protective effect on Na^+ observed in Fig. 3 is ascribed to a partial overlap of non-lipid and lipid events at the higher light intensity.) For G6P, on the other hand, disruption of the lipid bilayer, a lytic effect produced by peroxidative degradation of unsaturated lipids, appears to be necessary.

The occurrence of lipid peroxidation was initially assessed by the TBA reaction. As shown in Fig. 5, irradiation of PP-sensitized ghosts over a period long enough to cause >90% depletion of internal G6P resulted in a significant formation of TBA-reactive material. Chromogen production (ΔA_{532}) increased continuously and linearly with time, with no sign of a lag phase. In accordance with its effects on G6P release, BHT (0.5 mM) reduced the rate of lipid peroxidation by nearly 5-fold. Although quantitative, the TBA test has several shortcomings. First of all, it does not give a true picture of the overall status of membrane lipids and the protection provided by antioxidants, because non-propagated peroxides are not detected, including those that might arise via direct attack by oxygen in the singlet excited state ($^1O_2$) on unsaturated fatty acyl groups,

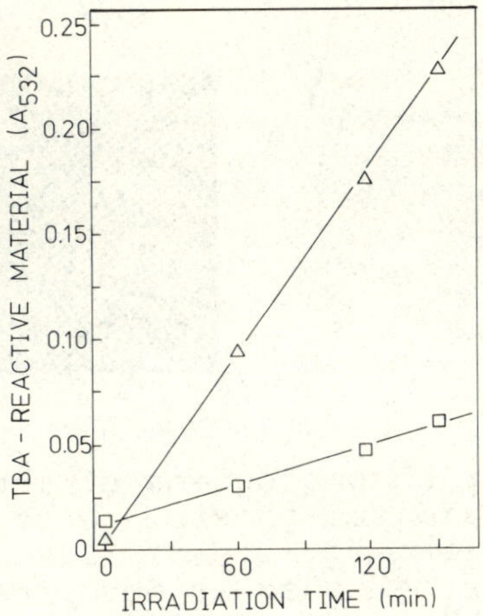

Figure 5. Photoperoxidation of membrane lipids as measured by the thiobarbituric acid (TBA) reaction. Resealed ghosts were irradiated in the absence (△) and presence (□) of 0.5 mM BHT. Light intensity was 25 W/m2 and PP concentration was 2 μM. Data is from the experiment shown in Fig. 4.

i.e. "ene reactions" (Foote, 1976). Secondly, peroxidation of cholesterol, which comprises about 25% of the RBC lipid mass, is not detected by this method. Accordingly, we adopted a more direct and informative approach, viz. TLC analysis of extracted lipid peroxides. As shown in Fig. 6 (left panel), PP-photosensitization produced peroxides of all the major lipid classes, including cholesterol, phosphatidylcholine, phosphatidylethanolamine, phosphatidylserine, and sphingomyelin. [Note that no peroxides could be detected in the dark controls (zero time)]. Our technique of spraying a common TLC plate with TMPD and DPH allows each peroxide(s) to be easily distinguished from its parent lipid, which migrates slightly faster (compare panels in Fig. 6). In agreement with the TBA evidence, peroxide formation occurred steadily throughout irradiation, with appreciable levels seen after 1 h. Peroxide spots were uniformly less intense in the BHT-treated ghosts, thereby confirming the results of the TBA analysis (Fig. 5). Protection against peroxidation was most obvious during the first 2 h of irradiation.

Differential permeability of the ghosts to Na^+ and G6P could also be demonstrated by a non-photochemical means, viz. by chemical modification. As shown in Fig. 7A, treatment of ghosts with PCMBS, which forms tight complexes with -SH groups, resulted in a rapid release of trapped Na^+. In-

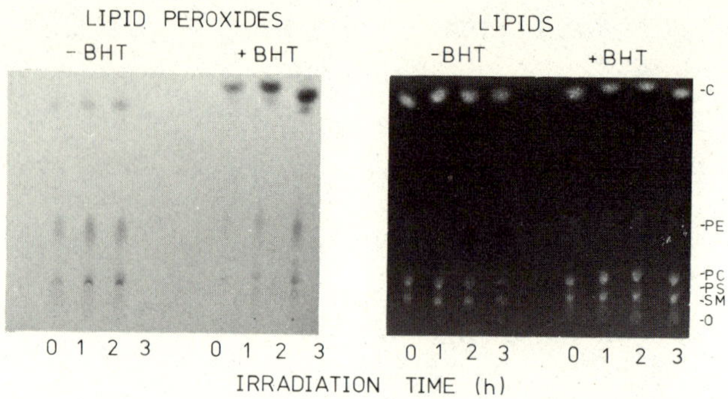

Figure 6. Thin layer chromatograms of membrane lipids from photoperoxidized ghosts. Photoreaction conditions are described in Fig. 3. Lipids were extracted and chromatographed as described in the Methods section. The major classes (C, Cholesterol; PE, phosphatidylethanolamine; PC, phosphatidylcholine; PS, phosphatidylserine; SM, sphingomyelin) were detected by DPH fluorescence (right panel). Lipid peroxides (left panel) were located by spraying with TMPD. The intense spot closest to the solvent front in the BHT-containing samples is an oxidation product of the antioxidant.

cubation with dithiothreitol prior to measuring Na^+ efflux substantially reversed this effect, indicating that the mercurial was not lysing the ghosts, but simply making them more leaky. These results are consistent with previous evidence (Grinstein and Rothstein, 1978) that modification of -SH groups on transmembrane regulators of cation permeability in RBCs results in accelerated cation leakage. Significantly, PCMBS treatment did not affect G6P efflux over an extended period of time (Fig. 7B). Thus, blocking -SH groups on specific membrane proteins mimics the early stages of the photoreaction, making the ghosts more permeable to Na^+, but not to G6P. We postulate that the -SH groups in question are highly susceptible to photooxidative attack, and that this produces the same effect as chemical blocking.

DISCUSSION

The results of this study demonstrate the usefulness of resealed RBC ghosts in investigating the membrane damaging effects of photodynamic sensitizers. Resealed ghosts (typically containing <5% of total protein as hemoglobin) have the following advantages over intact RBCs: a) Hemoglobin absorbs strongly in the λ max region of tetrapyrrole sensitizers, thereby interfering with their photoexcitation. b)

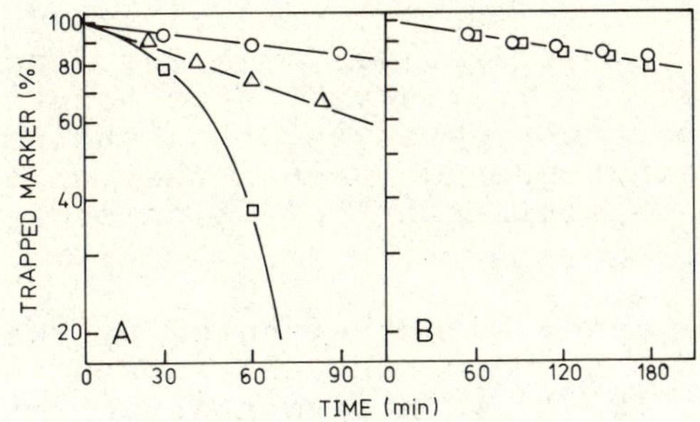

Figure 7. Effect of sulfhydryl modification on the permeability of resealed ghosts to Na^+ (panel A) and to G6P (panel B). Resealed ghosts (1 mg of protein/ml in PBS) were incubated with 200 μM PCMBS for 30 min in the dark (25°C). After rapid washing with PBC, marker efflux was monitored (□). Also shown is background efflux in untreated ghosts (o) and Na^+ release from PCMBS-treated ghosts that were washed first with PBS containing 5 mM dithiothreitol and then with PBC alone (△).

Intact RBCs scatter more light than resealed ghosts. c) Photooxidized membrane proteins and lipids can be analyzed directly without the inconvenience of (and potential artifacts produced by) removing hemoglobin. d) Lytic effects can still be examined, but without the complications of osmotic swelling. The absence of swelling under normal isotonic conditions has recently allowed us to determine the size range of the G6P-releasing lesions produced by photoactivated PP and bilirubin (Deziel and Girotti, 1982). Though resealed RBC ghosts may only approximate plasma membranes of other cells subjected to photodynamic attack, they are obviously superior to artificial bilayers (liposomes), for example, as experimental models for assessing the role of membrane damage in cell killing.

Membrane integrity of resealed ghosts irradiated in the presence of PP was monitored in terms of retention of trapped Na^+ and G6P. The choice of these particular markers and the use of relatively low light intensities and sensitizer concentrations was fortunate, since they allowed us to discriminate between two different release mechanisms. This distinction is based on the following key observations: a) The kinetic profile of Na^+ efflux differs from that of G6P. The cation emerged rapidly and continuously, and was almost completely released before the sugar phosphate began to emerge. Moreover, Na^+ continued to permeate at an accelerated rate after an abbreviated light dose, whereas G6P returned to its basal rate. b) The antioxidant BHT exerted significant protection against the light-induced efflux of G6P, but not Na^+. Qualitatively similar findings were reported earlier for bilirubin photosensitization (Deziel and Girotti, 1980a,b), thus verifying that photooxidative attack on the membrane is required, rather than some unique binding property of the sensitizer or insertion of sensitizer photoproducts.

That Na^+ release was not affected by BHT deserves additional comment. In previous related studies with unsealed ghosts (Girotti, 1979), we showed that antioxidants also had no effect on several manifestations of protein damage, e.g. spectrin cross-linking, oxidation of -SH groups, and inactivation of acetylcholinesterase, Na^+, K^+-ATPase and other enzymes. It appears to be more than just coincidence that Na^+ leakage and all of the protein-related effects examined to date are not inhibited by antioxidants. Rather, we suggest that direct photooxidative attack (non-lipid mediated) on known protein regulators of cation permeability (Grinstein and Rothstein, 1978; Huestis, 1977) is responsible for photoefflux of Na^+. The effects of the early photoreaction, rapid release of Na^+ before G6P, could be mimicked by chemically blocking specific -SH groups on these proteins (Fig. 7). An additional case for protein involvement is our previous evidence (Deziel and Girotti, 1980a,b) showing that antioxidants inhibited the release of both Na^+ and G6P from photosensitized liposomes which contained no proteins.

The fact that BHT protected against G6P efflux and lipid peroxidation (as measured by two independent methods) suggests that these effects are linked and that the antioxidant acted by intercepting lipid free radicals. We rule out quenching of excited state PP as the cause of BHT protection because other effects, e.g. Na^+ release, spectrin cross-linking, and oxidation of -SH groups, were not inhibited. Since BHT and various unsaturated fatty acids react with $^1O_2$ at comparable rates in organic solvents (Doleiden et al., 1974; Thomas and Foote, 1978), one may ask whether $^1O_2$ interception could explain the observed effects. We consider this unlikely for the following reason. From the known lipid composition of the RBC membrane (van Golde et al., 1967), we estimate that the suspension concentration of unsaturated lipids in our experiments was 1 mM. Therefore, simple competition for $^1O_2$ in the experiment shown in Fig. 6 would have required approx. 1 mM BHT for 50% inhibition. In actuality, an 80% inhibition was observed using 0.5 mM BHT (only part of which may have partitioned into the ghosts).

Both of the methods used to monitor lipid peroxidation indicated that this process occurs continuously throughout irradiation, showing none of the lag observed for G6P release. Thus this lag cannot be ascribed to a delay in the onset of lipid oxidation. Such a delay could conceivably be explained by oxidation of intrinsically more reactive amino acid residues or endogenous antioxidants before unsaturated lipids. We conclude, therefore, that an induction phase of lipid peroxidation is required for G6P to be released. Presumably lipid damage accumulates until a critical threshold is attained, after which the bilayer ruptures in a rapid, concerted fashion. We recently developed a technique for sizing these ruptures which is based on selective sedimentation in the presence of solutes of known molecular dimensions (Deziel and Girotti, 1982). The lesions ("pores") were estimated to be greater than 110 nm but less than 420 nm in diameter. This size range is less than 0.1% of the ghost's diameter, making it difficult, if not impossible, to detect these openings by other means, e.g. scanning electron microscopy. Therefore, these are not massive membrane lesions such as those produced by hypotonic lysis, but they nevertheless could be lethal to cells in which they occur.

SUMMARY

Photoactivation of protoporphyrin IX (PP) bound to resealed human erythrocyte (RBC) ghosts results in membrane damage which is manifested by the release of trapped markers Na^+ and glucose-6-phosphate (G6P). Efflux of Na^+ was rapid, continuous, and virtually complete before the onset of G6P efflux. The sugar phosphate emerged abruptly after a long lag. The antioxidant butylated hydroxytoluene (BHT) had no effect on the permeation of Na^+, but greatly suppressed that of G6P. These results suggest that the markers are emitted via different mechanisms. For G6P, disruption of the bilayer by free radical lipid peroxidation appears to be necessary, inasmuch as BHT inhibited peroxidation as measured by thiobarbituric acid reactivity and appearance of phospholipid and cholesterol hydroperoxides on thin layer chromatograms. It is deduced that non-lipid damage is sufficient for Na^+ release. This effect is manifested at low light intensities and low PP concentrations. Protein regulators of passive cation permeability may be the primary targets in this case. When sensitive sulfhydryl groups on these proteins were blocked with p-chloromercuri-benzenesulfonate, Na^+ leaked out rapidly, but G6P was unaffected, thereby mimicking the early stages of membrane photodamage.

ACKNOWLEDGEMENTS

This work was supported by Grant HD-07753 from the National Institutes of Health. M.R.D. is the recipient of the 1980-81 Walter H. Simson Memorial Fellowship in Photobiology from the Duro-Test Corporation, North Bergen, NJ.

REFERENCES

1. Asakawa, T., and Matsushita, S. (1980) Coloring Conditions of Thiobarbituric Acid Test for Detecting Lipid Hydroperoxides, Lipids 15, 137-140.

2. Deziel, M.R., and Girotti, A.W. (1980a) Bilirubin-Photosensitized Lysis of Resealed Erythrocyte Membranes, Photochem. Photobiol. 31, 593-596.

3. Deziel, M.R., and Girotti, A.W. (1980b) Photodynamic Action of Bilirubin on Liposomes and Erythrocyte Membranes, J. Biol. Chem. 255, 8192-8198.

4. Deziel, M.R., and Girotti, A.W. (1982) Lysis of Resealed Erythrocyte Ghosts by Photoactivated Tetrapyrroles: Estimation of Photolesion Dimensions, Int. J. Biochem. (in press).

5. Doleiden, F.H., Farenholtz, S.R., Lamola, A.A., and Trozzolo, A.M. (1974) Reactivity of Cholesterol and Some Fatty Acids Toward Singlet Oxygen, Photochem. Photobiol. 20, 519-521.

6. Dougherty, T.J., Grindley, G.B., Fiel, R., Weishaupt, K.R., and Boyle, D.G. (1975) Photoradiation Therapy II. Cure of Animal Tumors with Hematoporphyrin and Light, J. Natl. Cancer Inst. 55, 115-120.

7. Foote, C.S. (1976) Photosensitized Oxygenation and Singlet Oxygen: Consequences in Biological Systems, in Pryor, W.A. (ed.): Free Radicals in Biology, Academic Press, New York, Vol. 2, pp. 85-133.

8. Girotti, A.W. (1979) Protoporphyrin-Sensitized Photodamage in Isolated Membranes of Human Erythrocytes, Biochem. 18, 4403-4411.

9. Grinstein, S., and Rothstein, A. (1978) Chemically Induced Cation Permeability in Red Cell Membrane Vesicles: The Sidedness of the Response and the Proteins Involved, Biochim. Biophys. Acta 508, 236-245.

10. Huestis, W.H. (1977) A Sodium-Specific Membrane Permeability Defect Induced by Phospholipid Vesicle Treatment of Erythrocytes, J. Biol. Chem. 252, 6764-6768.

11. Hyslop, P.A. and York, D.A. (1980) The Use of 1,6-Diphenylhexatriene to Detect Lipids on Thin-Layer Chromatograms, Anal. Biochem. 101, 75-77.

12. Kessel, D. (1977) Effects of Photoactivated Porphyrins at the Cell Surface of Leukemic L1210 Cells, Biochem.

16, 3443-3449.

13. Kohn, K., and Kessel, D., (1979) On the Mode of Cytotoxic Action of Photoactivated Porphyrins, Biochem. Pharmacol. 28, 2465-2470.

14. Mauk, M.R., and Girotti, A.W. (1973) Photooxidation of the Protoporphyrin-Apomyoglobin Complex, Biochem. 12, 3187-3193.

15. Smith, L.L., and Hill, F.L. (1972) Detection of Sterol Hydroperoxides on Thin-Layer Chromatoplates by Means of the Wurster Dyes, J. Chromatogr. 66, 101-109.

16. Thomas, M.J., and Foote, C.S. (1978) Chemistry of Singlet Oxygen XXVI. Photooxygenation of Phenols, Photochem. Photobiol. 27, 683-693.

17. van Golde, L.M.G., Tomasi, V., and van Deenen, L.L.M. (1967) Determination of Molecular Species of Lecithin from Erythrocytes and Plasma, Chem. Phys. Lipids 1, 282-293.

PHOTODYNAMIC MEMBRANE DAMAGE

J. van Steveninck, T.M.A.R. Dubbelman and H. Verweij

Sylvius Laboratories
Department of Medical Biochemistry
Wassenaarseweg 72
2333 AL Leiden, The Netherlands

INTRODUCTION

The use of porphyrins in photochemotherapy of neoplastic disease is based on the formation of singlet oxygen (Spikes, 1975; Weishaupt et al., 1976; Kessel, 1981) and possibly hydroxyl radicals (Hariharan et al., 1980), subsequent to photoactivation of the sensitizer. The mechanism of cell killing is still unknown. Many studies indicate that membrane and DNA damage are involved, but the relative contribution of these processes to cell inactivation is still a matter of dispute.

Previous studies have shown that from a biochemical point of view membrane damage is characterized by photooxidation of sensitive amino acid residues in membrane proteins (Dubbelman et al., 1978a), cross-linking of membrane proteins (Dubbelman et al., 1978b; Girotti, 1975, 1980) and mutilation of ultrastructure (De Goeij et al., 1975). Although dissolved unsaturated fatty acid and cholesterol are highly sensitive to photooxidation (Doleiden et al., 1974), it has been shown that these compounds are not photooxidized in intact membranes to an appreciable extent and that their very slow photooxidation under these circumstances does not contribute to impairment of membrane function (De Goeij et al., 1976; De Goeij and Van Steveninck, 1976). Membrane damage in the sense of disturbed membrane functions is caused solely by photodynamic effects on membrane proteins.

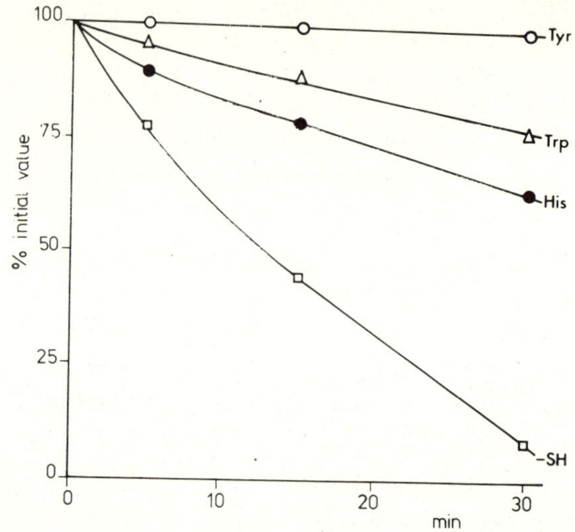

Figure 1. Photooxidation of cysteine (□), histidine (•), tryptophan (Δ) and tyrosine (o) in spectrin. A solution of spectrin (0.7 mg/ml) was illuminated in 10 mM phosphate buffer, pH 8.0, in the presence of 0.05 mM protoporphyrin.

PHOTODYNAMIC EFFECTS ON MEMBRANE PROTEINS

The primary effect of illumination of proteins in the presence of a sensitizer is photooxidation of sensitive amino acid residues. In the pH range 7.0-8.5, the sulfhydryl groups in spectrin are most sensitive to photooxidation followed by histidine, tryptophan and tyrosine (Fig. 1). The same sequence of sensitivity was found with other proteins, like band 3 of the red blood cell membrane and myoglobin.

In many cases, protein photooxidation is followed by a secondary covalent cross-linking. Interpeptide cross-linking can be visualized by sodium dodecylsulfate polyacrylamide gel electrophoresis, as shown in Fig. 2. During illumination in the presence of a sensitizer, protein bands disappear from their normal position in the electropherogram with a concomitant accumulation of high-molecular weight cross-linked material on top of the gels.

Photooxidation and secondary cross-linking are not obligatory coupled. This is affirmed by the following experimental observations.

1. In the temperature range from 0° to 40°C, photooxidation

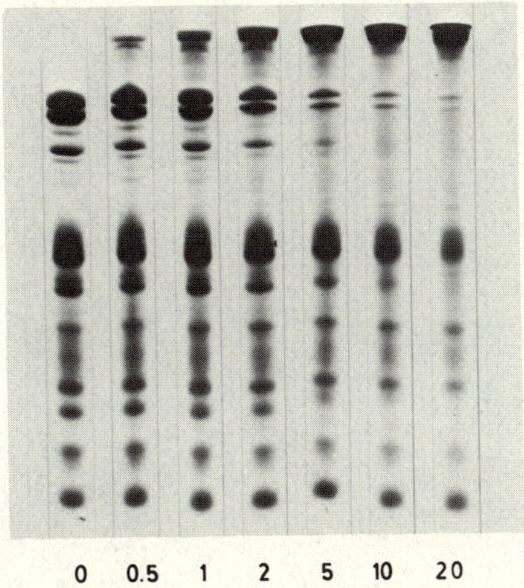

Figure 2. Photodynamic cross-linking of red cell membrane proteins as shown by SDS-polyacrylamide gel electro-phoresis. Erythrocyte membranes (2 mg protein/ml) were illuminated in 10 mM phosphate buffer, pH 8.0, in the presence of 0.05 mM protoporphyrin, during 0-20 minutes.

is virtually temperature independent but the extent of interpeptide cross-linking is decreased significantly at lower temperatures (Fig. 3). Apparently cross-linking has a much higher energy of activation than does photooxidation (Dubbelman et al., 1980a).

2. Some proteins, like myoglobin, are photooxidized readily without exhibiting interpeptide cross-linking, as judged from polyacrylamide gel electrophoresis (Girotti et al., 1979). In the presence of 8 M urea, however, photooxidation of myoglobin is not affected, whereas polyacrylamide gel electrophoresis clearly shows covalent cross-linking under these experimental conditions (Fig. 4).

3. Amidination of NH_2 groups causes inhibition of cross-linking in certain proteins, as will be discussed in more detail in the next section. Photooxidation, on the other hand, is not inhibited by amidination (Fig. 4).

4. When a protein sample, illuminated in the presence of a sensitizer, is put in the dark, photooxidation stops immediately. Under appropriate experimental conditions it can be shown that cross-linking proceeds in the dark(Dubbelman et al, 1980a).

All these observations are in accordance with the notion that cross-linking succeeds photooxidation as a secondary, but not compulsory reaction. This cross-linking must

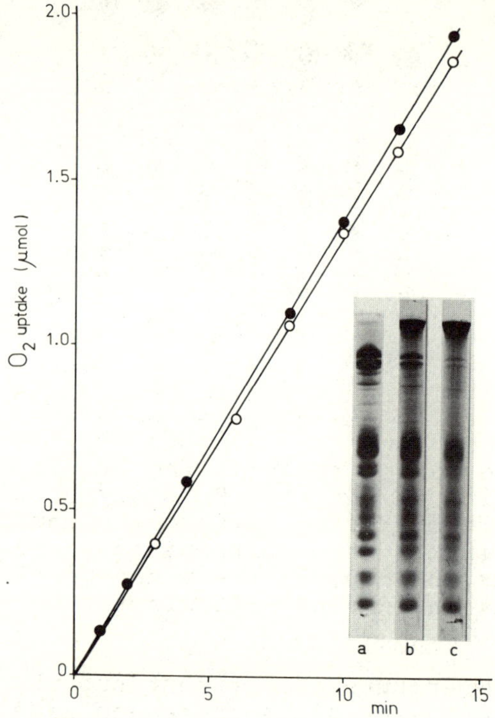

Figure 3. Oxygen consumption during illumination of 3 ml red cell membrane suspension (2 mg protein/ml) in 10 mM phosphate buffer, pH 8.0, containing 0.05 mM protoporphyrin, at 0°C (o) and 37°C (•). Inset: electropherogram of a: dark control; b: sample, illuminated during 15 minutes at 0°C; c: sample, illuminated during 15 minutes at 37°C.

be caused by an interaction between photooxidized amino acid residues and reactive groups in the protein molecule, leading to the formation of a covalent bond. Several experimental approaches have been used to study the chemical nature of photodynamic cross-links.

a) Analysis with group-specific reagents.

During illumination of proteins in the presence of a sensitizer, a decrease of amino groups is observed. As NH_2 groups are not sensitive to photooxidation, these results indicate a secondary cross-linking reaction between photooxidized amino acid residues and amino groups (Dubbelman et al, 1978a; Verweij et al, 1981). This interpretation is strongly supported by previous observation of Tomita et al. (1969) and Straight and Spikes (1978). Analyzing photo-

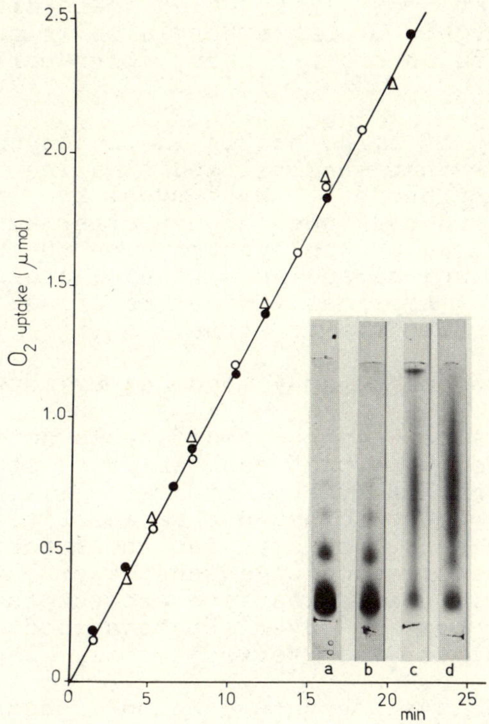

Figure 4. Oxygen consumption during illumination of 3 ml myoglobin (1 mg/ml) at pH 8.0 in the presence of 0.05 mM protoporphyrin. o: in 10 mM phosphate buffer; •: in 10 mM phosphate buffer, containing 8 M urea; △: in 10 mM phosphate buffer, containing 8 M urea, after amidination of the protein. Inset: electropherogram of a: dark control; b: sample, illuminated during 20 minutes in buffer; c: illuminated during 20 minutes in buffer containing 8 M urea; d: as c, after amidination of the protein.

oxidation products of histidine, Tomita et al. elucidated the reaction between the oxidized imidazole group and the amino group of the same, or a different histidine molecule. Straight and Spikes described the decrease of NH_2 groups during photosensitized oxidation of histidine, tryptophan and tyrosine.

An interesting observation is that decrease of NH_2 groups is also observed during photooxidation of proteins which do not show interpeptide cross-linking on polyacrylamide gel electropherograms, like native myoglobin. The only obvious interpretation is the involvement of NH_2 groups in intramolecular cross-links under these circumstances. This makes it highly probable that also in proteins, exhibiting intermolecular cross-linking, intramolecular cross-links will occur.

Involvement of amino groups in photodynamic cross-linking suggests that chemical modification of NH_2 groups might inhibit the process. With myoglobin in 8 M urea a partial inhibition was indeed found after amidination of the NH_2 groups (Fig. 4). With spectrin, on the other hand, amidination did not cause inhibition of cross-linking (Girotti, 1980). Apparently other reactive groups may also be involved in photodynamic cross-linking.

b) Thin layer chromatography model experiments.

Cross-links between photooxidizable amino acid residues and various reactive groups can easily be studied in model experiments. In this type of studies a mixture of a photooxidizable amino acid and a radioactively labeled, non-photooxidizable substance carrying the reactive group is illuminated in the presence of a sensitizer. A cross-linking reaction will yield a radioactive product that can be visualized e.g. by thin layer chromatography of the mixture (Dubbelman et al., 1978a; Verweij et al., 1981).

c) Model studies with sepharose-bound ligands.

With this approach, the compound carrying the reactive group is coupled to activated sepharose. This gel is mixed with a solution of a radioactively labeled, photooxidizable amino acid and illuminated in the presence of a sensitizer. Cross-linking is reflected by covalent binding of radioactive label to the gel (Verweij and Van Steveninck, 1981).

Fig. 5 summarizes the types of cross-links found with these various studies. Conspicuous is the predominate role of histidine. Together with the high sensitivity of histidine to photooxidation this suggests that photooxidized histidine will be involved in the majority of intra- and interpeptide photodynamic cross-links.

PHOTODYNAMIC DETERIORATION OF MEMBRANE FUNCTIONS.

An intriguing question is whether deterioration of membrane function is caused by photooxidation of essential amino acid residues, intrapeptide cross-linking or interpeptide cross-linking. Only in a few cases can a tentative answer, based on experimental observations, be given.

Exposure of human erythrocytes to light in the presence of a sensitizer leads very quickly to a decrease of cell deformability (Dubbelman et al., 1977). As spectrin plays a

	photo-oxidized				
cross-link to:	His	Trp	Tyr	Cys	Met
$-NH_2$	++	+	+	–	–
$-SH$	++	+	+/–	++	–
$-NH-C(=NH)-NH_2$	+/–	+/–	+/–	–	–
$-S-CH_3$	–	–	–	–	–
$-OH$	–	–	–	–	–
$-C(=O)-NH_2$	–	–	–	–	–
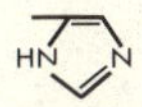	++	++	+	–	–
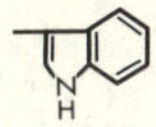	++	–	–	–	–
	+	–	–	–	–

Figure 5. Photodynamic cross-linking of photooxidized histidine, tryptophan, tyrosine, cysteine and methionine to various reactive groups. ++, +, +/- and - refer to strong, moderate, weak and no cross-linking. The cysteine-cysteine cross-link is completely disrupted by dithiothreitol.

predominant role in the regulation of the visco-elastic properties of the red blood cell, it is highly probable that decreased cell deformability is attributable to photodynamic effects on this protein. Presumably interpeptide cross-linking rather than photooxidation is the decisive process in this case. This supposition is based on two observations. First, there was a very close correlation between the extent of decreased cell deformability and spectrin cross-linking. Further, if a similar degree of spectrin cross-linking was provoked by other means, e.g. by treatment with bifunctional cross-linking agents, a similar decrease of cell deformability was observed (Dubbelman et al., 1977; Fischer et al., 1978). The supposition that spectrin cross-linking is responsible for the decreased cell deformability has a certain similarity to the proposed mechanism of deformability control by spectrin. Variable associations of spectrin molecules with each other constitute an essential component of the physiological control mechanism (Marchesi, 1979).

Photohemolysis, subsequent to illumination of red blood cells in the presence of a sensitizer, is a colloid-osmotic process, based on increased cation permeability of the membrane (Schothorst et al., 1970). In contrast to decreased cell deformability, this increased cation permeability seems to be caused by photooxidation of essential amino acid residues rather than by cross-linking of membrane proteins (Dubbelman et al., 1980a). This was deduced from the fact that K+ leakage from the cells is virtually temperature independent, like photooxidation, whereas cross-linking has a strong temperature dependency (Fig. 3).

Band 6 protein of human erythrocyte membranes, representing the monomeric form of glyceraldehyde-3-phosphate dehydrogenase, is very sensitive to photodynamic inactivation. Several lines of evidence indicate that loss of enzyme activity is directly correlated to photooxidation of the essential thiol group in the active centre of the enzyme (Dubbelman et al., 1978b).

Many carrier-mediated transport processes in biomembranes are inhibited by photosensitized processes (Kessel, 1977; Dubbelman et al.,1980b). In a recent study the background of photodynamic inhibition of anion transport was studied in detail (Dubbelman et al., 1981). Anion transport in human erythrocytes is mediated by band 3 protein (Ramjeesingh et al., 1980; Jennings and Passow, 1979). Detailed studies of band 3 protein have shown that this polypeptide traverses the lipid bilayer of the membrane at least three times (Fig. 6). Band 3 can be fragmented in situ by controlled peptic digestion (Jennings and Passow, 1979; Ramjeesingh et al., 1980). With this technique, the polypeptide can be subdivided in an extracellular, a transmembrane and an intracellular fragment (Fig. 6). The extracellular fragment is the C-terminal portion of the protein, has a molecular weight of about 35K daltons, and carries the carbohydrate portion of the molecule. The transmembrane fragment is relatively hydrophobic, has a molecular weight of about 17K daltons and crosses the membrane three times. The intracellular fragment has a molecular weight of about 40K daltons and represents the N-terminal portion of the molecule.

4,4'-Diisothiocyano dihydrostilbene-2,2'-disulfonic acid (H_2DIDS) is a very strong inhibitor of anion transport. It appeared that complete inhibition of transport is correlated with the binding of one H_2DIDS molecule per band 3 monomer (Lepke et al., 1976). It has been shown that the H_2DIDS binding site is located in the extracellular loop of the 17K transmembrane fragment (Fig. 6). Detailed studies of protoporphyrin-induced photodynamic effects on band 3 protein yielded the following results (Dubbelman et al., 1981).

1. During illumination in the presence of protoporphyrin, photooxidizable amino acid residues were destroyed with a sequence of decreasing sensitivity of SH > histidine > tryptophan > tyrosine. A concomitant decrease of NH_2 groups was observed.

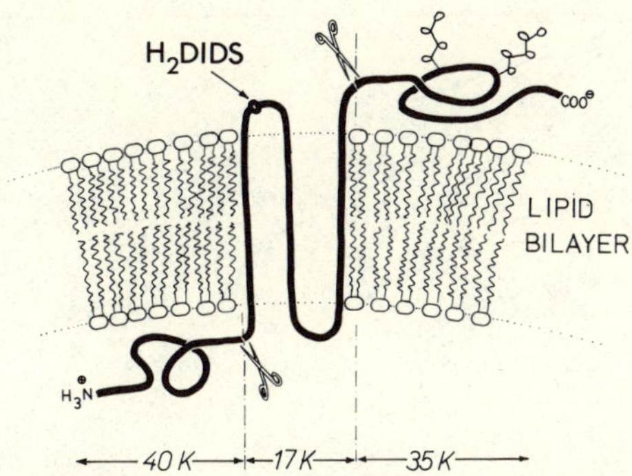

Figure 6. Arrangement of band 3 protein in the red blood cell membrane, showing the intracellular, the transmembrane and the extracellular fragments.

2. Both intact band 3, the extracellular fragment and the intracellular fragment were cross-linked, whereas the transmembrane fragment was completely insensitive to photodynamic cross-linking.

3. Anion transport was progressively inhibited. This inhibition proceeded much faster than interpeptide cross-linking.

4. Subsequent to illumination H_2DIDS binding to band 3 protein was decreased. There was a close parallel between transport inhibition and decreased H_2DIDS binding (Fig. 7).

5. With deuteroporphyrin 2,4-disulfonic acid instead of protoporphyrin a similar correlation between anion transport inhibition and decreased H_2DIDS binding was found, but with tnis sensitizer no band 3 interpeptide cross-linking occurred (Fig.8).

These results clearly show that photodynamic inhibition of anion transport is not caused by interpeptide cross-linking of the carrier, but by photooxidation at or near the active site, localized in the not-cross-linking transmembrane fragment of band 3 protein. More difficult to decide is whether photooxidation of an essential amino acid residue or intrapeptide cross-linking is the direct cause of transport inhibition. In this context, it is interesting that the H_2DIDS binding site has recently been identified as a lysine residue (Ramjeesingh et al., 1981). Decreased H_2DIDS binding can therefore be explained either via perturbation of the local conformation around the H_2DIDS binding site, thus hindering binding, or via intrapeptide cross-linking, involving the H_2DIDS binding amino group.

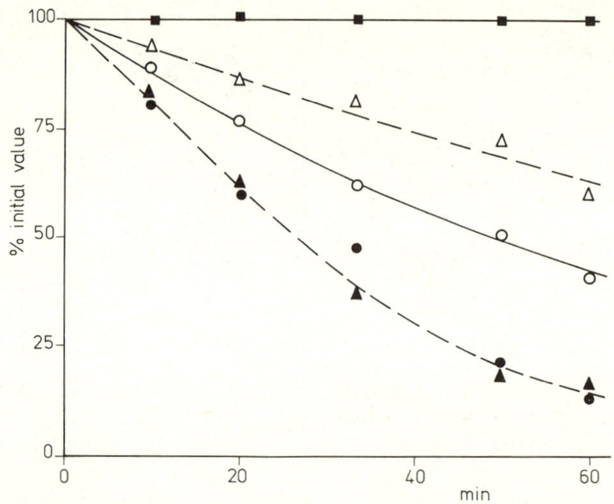

Figure 7. Photodynamic effects on band 3 protein and spectrin in intact red blood cells. Erythrocytes were illuminated in 1% NaCl at pH 7.5 in the presence of 0.2 mM protoporphyrin. ▲-▲: sulfate transport velocity; •-•: H_2DIDS binding subsequent to illumination during the indicated period. o-o: spectrin decrease, due to cross-linking; △-△: band 3 decrease due to cross-linking; ■-■: 17K fragment.

CONCLUSION AND PERSPECTIVES

Photodynamic membrane damage is caused primarily by photooxidation of membrane proteins. Theoretically both photooxidation of essential amino acid residues, intrapeptide cross-linking and interpeptide cross-linking can be involved in deterioration of protein function. It was possible in one case to ascribe disturbed protein function with reasonable probability to interpeptide cross-linking: decreased red cell deformability seems to be caused by interpeptide spectrin cross-linking.

In general, intramolecular photodynamic processes seem to be more important with respect to disturbed function. Until recently, the only intramolecular process that was considered in this context was photooxidation of essential amino acid residues. As indicated above, the possible significance of intrapeptide cross-links may have been underestimated previously. In future investigations the consequences of these intrapeptide cross-links will have to be

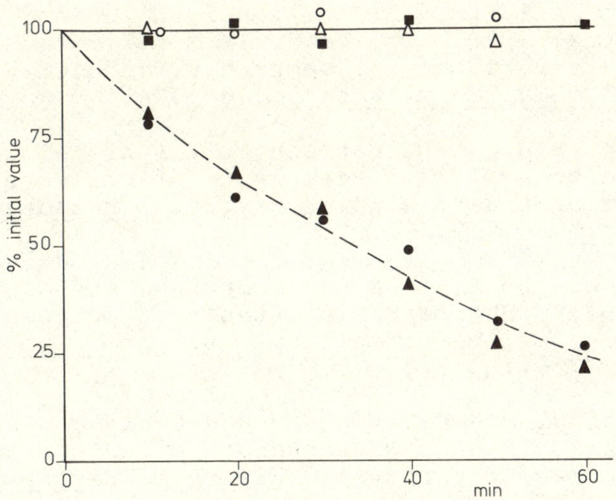

Figure 8. Identical experiment as shown in Fig. 7, but with 0.8 mM deuteroporphyrin 2,4-disulfonic acid as sensitizer. Symbols: see legend to Fig. 7.

studied in more detail.

An important observation is that during photodynamic anion transport inhibition a considerable cross-linking of band 3 and especially of spectrin is found with protoporphyrin as sensitizer, but not with deuterporphyrin 2,4-disulfonic acid. Apparently deuteroporphyrin 2,4-disulfonic acid has a certain degree of specificity with respect to photodynamic inhibition of anion transport. This type of specificity may be of interest in the development of new sensitizers for phototherapy of specific targets, e.g., the Na^+/K^+ pump. Further studies will be needed to explore these possibilities.

REFERENCES

1. De Goeij, A.F.P.M., P.H.J.T. Ververgaert and J. van Steveninck, 1975. Photodynamic effects of protoporphyrin on the architecture of erythrocyte membranes in protoporphyria and in normal red blood cells. Clin. Chim. Acta 62:287-292.

2. De Goeij, A.F.P.M. and J. van Steveninck, 1976. Photodynamic effects of protoporphyrin on cholesterol and unsaturated fatty acids in erythrocyte membranes in protoporphyria and in normal red blood cells. Clin. Chim. Acta 68:115-122.

3. De Goeij, A.F.P.M., R.J.C. van Straalen and J. van Steveninck, 1976. Photodynamic modification of proteins in human red blood cell membranes, induced by protoporphyrin. Clin. Chim. Acta 71:485-494.

4. Doleiden, F.H., S.R. Fahrenholtz, A.A. Lamola and A.M. Trozzolo, 1974. Reactivity of cholesterol and some fatty acids toward singlet oxygen. Photochem. Photobiol. 20:519-521.

5. Dubbelman, T.M.A.R., A.W. de Bruijne and J. van Steveninck, 1977. Photodynamic effects of protoporphyrin on red blood cell deformability. Biochem. Biophys. Res. Commun. 77:811-817.

6. Dubbelman, T.M.A.R., A.F.P.M. de Goeij and J. van Steveninck, 1978a. Photodynamic effects of protoporphyrin on human erythrocytes. Biochim. Biophys. Acta 511:141-151.

7. Dubbelman, T.M.A.R., A.F.P.M. de Goeij and J. van Steveninck, 1978b. Protoporphyrin-sensitized photodynamic modification of proteins in isolated human red blood cell membranes. Photochem. Photobiol. 28:197-204.

8. Dubbelman, T.M.A.R., C. Haasnoot and J. van Steveninck, 1980a. Temperature dependence of photodynamic red cell membrane damage. Biochim. Biophys. Acta 601:220-227.

9. Dubbelman, T.M.A.R., A.F.P.M. de Goeij and J. van Steveninck, 1980b. Protoporphyrin-induced photodynamic effects on transport processes across the membrane of human erythrocytes. Biochim. Biophys.Acta 595:133-139.

10. Dubbelman, T.M.A.R., A.F.P.M. de Goeij, K. Christianse and J. van Steveninck, 1981. Protoporphyrin-induced photodynamic effects on band 3 protein of human erythrocyte membranes. Biochim. Biophys. Acta 649: 310-316.

11. Fischer, T.M., C.W.M. Haest, M. Stöhr, D. Kamp and B. Deuticke, 1978. Selective alteration of erythrocyte deformability by SH-reagents. Evidence for an involvement of spectrin in membrane shear elasticity. Biochim. Biophys. Acta 510:270-282.

12. Girotti, A.W., 1975. Photodynamic action of bilirubin on human erythrocyte membranes. Modification of polypeptide constituents. Biochem. 14:3377-3383.

13. Girotti. A.W., S. Lyman and M.R. Deziel, 1979. Methylene blue-sensitized photooxidation of hemoglobin: evidence for cross-link formation. Photochem.

Photobiol. 29:1119-1125.

14. Girotti, A.W., 1980. Photosensitized cross-linking of erythrocyte membrane proteins. Biochim. Biophys. Acta 602:45-56.

15. Hariharan, P.V., J. Courtney and S. Eleczko, 1980. Production of hydroxyl radicals in cell systems exposed to haematoporphyrin and red light. Int. J. Radiat. Biol. 37:691-694.

16. Jennings, M.L. and H. Passow, 1979. Anion transport across the erythrocyte membrane, in situ proteolysis of band 3 protein and cross-linking of proteolytic fragments by 4,4'-diisothiocyano dihydrostilbene-2,2'-disulfonate. Biochim. Biophys. Acta 554:498-519.

17. Kessel, D., 1977. Effects of photoactiviated porphyrins at the cell surface of leukemia L1210 cells. Biochem. 16:3443-3449.

18. Kessel, D., 1981. Transport and binding of hematoporphyrin derivative and related porphyrins by Murine Leukemia L1210 cells. Cancer Res. 41:1318-1323.

19. Lepke, S., H. Fasold, M. Pring and H. Passow, 1976. A study of the relationship between inhibition of anion exchange and binding to the red blood cell membrane of DIDS and H_2DIDS. J. Membrane Biol. 29:147-177.

20. Marchesi, V.T. 1979. Spectrin: present status of a putative cyto-skeletal Protein of the red cell membrane. J. Membrane Biol. 51:101-131.

21. Ramjeesingh, M., A. Gaarn and A. Rothstein, 1980. The location of a disulfonic stilbene binding site in band 3, the anion transport protein of the red blood cell membrane. Biochim. Biophys. Acta 599:127-139.

22. Ramjeesingh, M., A. Gaarn and A. Rothstein, 1981. The amino acid conjugate formed by the interaction of the anion transport inhibitor 4,4'-diisothiocyano-2,2'-stilbenedisulfonic acid with band 3 protein from human red blood cell membranes. Biochim. Biophys. Acta 641:173-182.

23. Schothorst, A.A., J. van Steveninck, L.N. Went and D. Suurmond, 1970. Protoporphyrin-induced photohemolysis in protoporphyria and in normal red blood cells. Clin. Chim. Acta 28:41-49.

24. Spector, M., S. O'Neal and E. Racker, 1980. Reconstitution of the Na^+K^+ pump of Ehrlich Ascites Tumor and en-

hancement of efficiency by quercetin. J. Biol. Chem. 255:5504-5507.

25. Spikes, J.D., 1975. Porphyrins and related compounds as photodynamic sensitizers. Ann. N. Y. Acad. Sci. 244:496-508.

26. Straight, R. and J.D. Spikes, 1978. Sensitized photooxidation of amino acids: effects on the reactivity of their primary amino groups with fluorescamine and o-phthalaldehyde. Photochem. Photobiol. 27:565-569.

27. Tomita, M., M. Irie and T. Ukita, 1969. Sensitized photooxidation of histidine and its derivatives. Products and mechanism of the reaction. Biochem. 8: 5149-5160.

28. Verweij, H., T.M.A.R. Dubbelman and J. van Steveninck, 1981. Photodynamic protein cross-linking. Biochim. Biophys. Acta 647:87-94.

29. Verweij, H. and J. van Steveninck, 1981. Model studies on photodynamic cross-linking. Photochem. Photobiol., 35:265-267.

30. Weishaupt, K.R., C.J. Gomer and T.J. Dougherty, 1976. Identification of singlet oxygen as the cytotoxic agent in photodestruction of a murine tumor. Cancer Res. 36: 2326-2329.

HPD - A STUDY OF ITS COMPONENTS AND THEIR PROPERTIES

R. Bonnett

Department of Chemistry
Queen Mary College
Mile End Road
London E1 4NS

M.C. Berenbaum

Wellcome Laboratories of Experimental Pathology
Variety Club Research Wing
St. Mary's Hospital Medical School
London W2 1PG

The phototoxic effect ("photodynamic action") of administered porphyrin in mammals appears to have been first observed by Hausmann (1) in mice and by Meyer-Betz (2) in man. The effect was known in certain porphyrias long before this, but the cause was not known. The localisation of administered porphyrins in tumour tissue was recognised (3) in the 1940s, and this behaviour was explored further in the 1950s in work that led to the development of the HPD preparation (4,5). It was not until 1972 that these two ideas concerning the biological activity of porphyrins (that is, photodegradation of tissue and localisation in tumours) came together successfully, when Diamond and his colleagues (6) demonstrated that a porphyrin could be used to photosensitise the preferential degradation of tumour implants in the rat. In 1975 this result was confirmed and extended by Berenbaum and his colleagues (7), and by Dougherty and his colleagues (8).

We have been working on this problem since 1976 with support from the Medical Research Council. Our approach has been to try to isolate the various fractions or constituents of HPD, and to characterise the constituents which could be

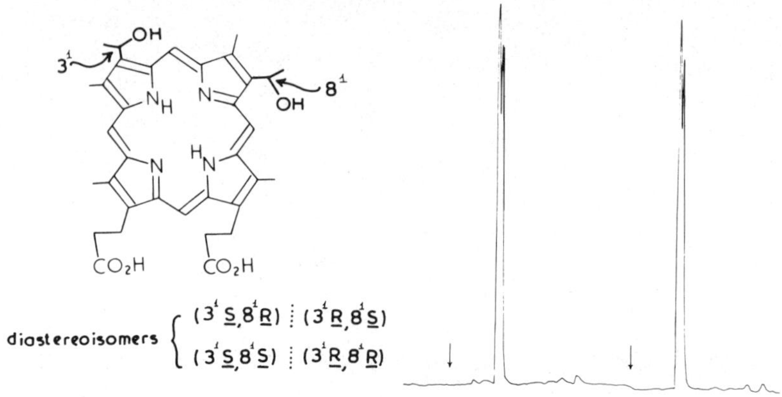

Figure 1. Analytical HPLC of haematoporphyrin (supplied as the dihydrochloride) with resolution of the two diastereoisomers of structures shown and showing the reproducibility of the separation. (μ-Bondapak C18, $MeOH:H_2O=4:1$, 1 mM in $Bu_4NH_2PO_4$)

obtained pure. Part of the characterisation (photophysical parameters) has involved collaboration with Professor T.G. Truscott and his colleagues at Paisley College of Technology (laser flash photolysis) and Dr. E.J. Land at the Christie Hospital, Manchester (pulse radiolysis). On the biological side we decided to avoid fluorescence work, localisation studies, and <u>in vitro</u> biological work, and to measure the photonecrosis of tumour implants in mice.

Haematoporphyrin derivative is made by treating haematoporphyrin with sulphuric acid in acetic acid at room temperature[1]. We have analysed it using the reverse-phase HPLC method (9). We introduced this analytical method for porphyrin carboxylic acids in 1978, when HPD was given as one of several examples of its application. It was clear then that HPD was a complex mixture of which the major components were the mono- and di-acetates of haematoporphyrin.

Haematoporphyrin, as commercially supplied as the dihydrochloride, is sometimes itself grossly contaminated. We have worked with a batch of purity >95%: the chromatogram of this material is shown in Fig. 1. The bifurcation of the peak is due to the two diastereoisomers of haematoporphyrin. Haematoporphyrin has two different chiral centres at 3^1 and

---

[1]We use HPD to indicate the solid product of this reaction. Alkali-treated HPD is chemically quite distinct (see later).

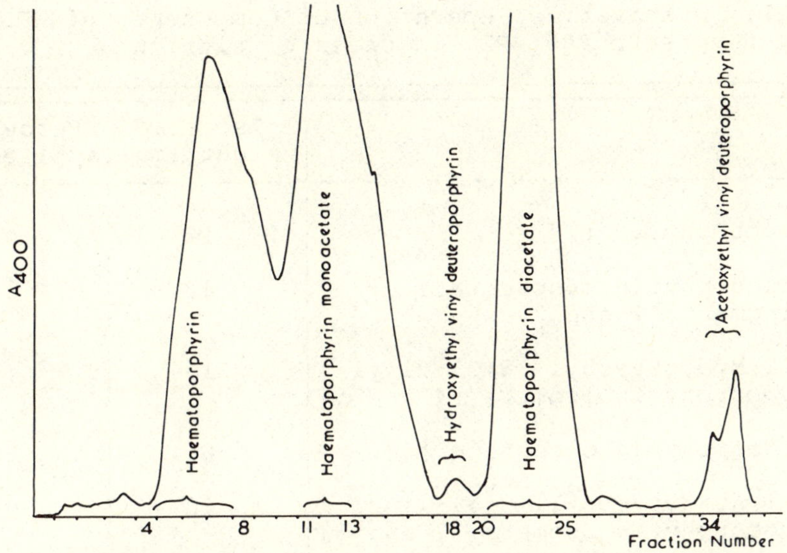

Figure 2. Preparative separation of the components of HPD on reverse phase HPLC. (200 g Merck Lichroprep RP-18 eluted with methanol:water containing 4% acetic acid. The proportion of methanol is increased in small steps during the elution) (11).

8^1 (Fig. 1) and hence there are two diastereoisomers (and four enantiomers). The two diastereoisomers have been separated by preparative HPLC, and both the diacids and the dimethyl esters have been characterised. Attempts to carry out X-ray structure analysis have been frustrated so far by the poor quality of the crystals, so we have two substances, and two isomeric structures, but we do not at present know which corresponds to which. We designate that haematoporphyrin isomer which comes off the reverse phase column first as α, the other β. Since stereochemistry is usually a significant factor in determining biological activity, this seemingly minor and uninvestigated aspect of haematoporphyrin structure might in the longer term turn out to be quite important.

Scaling up the analytical HPLC separation of HPD (9) gave rise to certain problems because the haematoporphyrin acetates, being pseudo benzylic acetates, are rather reactive. Separation has to be carried out expeditiously, and the products need to be worked up without delay. Here the skill and tenacity of Peter Scourides played a large part in ensuring a successful outcome. The preparative separations were made on a column of Merck Lichroprep RP-18 reverse

Table 1. Preparative Separation of Components of HPD (294 mg) by reverse phase HPLC in order of elution

	Isolated weight (mg)	Percent of applied HPD
Haematoporphyrin [1]	12	4
Haematoporphyrin monoacetate [two isomers, 2 and 3]	42	14
3(8)-(1-Hydroxyethyl)-8(3)-vinyl deuteroporphyrin isomers [4 and 5]	13	4
Haematoporphyrin diacetate [6]	66	22
3(8)-(1-Acetoxyethyl)-8(3)-vinyl deuteroporphyrin isomers [7 and 8]	20	6
Protoporphyrin [9]	*	

* Not obtained as a separate component from the preparation column.

phase silica gel (25-40 μm, 200 g, column diameter 40 mm) using the Jobin-Yvon Chromatospac 10 eluted with methanol: water containing 4% acetic acid. As the fractions emerged they were examined by analytical HPLC. Appropriate fractions were combined to give the products as the dicarboxylic acids, which were identified by spectroscopic methods and by comparison with authentic compounds prepared by standard routes (10). The latter are now the preferred routes to the various components (11). Our separation thus led straight to the dicarboxylic acids, which were needed for biological testing: Clezy and his colleagues have described an independent approach based on the separation of esters by column chromatography (12).

A typical preparative chromatogram is shown in Fig. 2 (note that absorbance is not corrected for ε_{400} of the various components, so that the areas do not correspond to molar proportions. This is important because the Soret bands are sharp, and ε in this region changes rapidly with wavelength and structure). The amounts and identities of the compounds isolated as amorphous solids from 294 mg of HPD are shown in Table 1. Structures are given in Fig. 3.

We thus find that HPD is a complex mixture, the con-

Table 2. Data for triplet species in acetone (17)

Compound	First order rate constant for decay k_1 (s^{-1})	Second order rate constant for quenching by oxygen k_Q ($M^{-1}s^{-1}$)
Haematoporphyrin [1]	4.0×10^4	1.4×10^9
Haematoporphyrin α*	4.1×10^4	not measured
Haematoporphyrin β	3.6×10^4	1.3×10^9
Haematoporphyrin monoacetate [2,3]	6.1×10^4	1.5×10^9
Haematoporphyrin diacetate [6]	12.0×10^4	1.7×10^9

* Refers to that diastereoisomer of haematoporphyrin which emerges first from reverse phase column.

stitution of which may vary from one preparation to another. The main components in our preparation are always the diacetate [6] and the monoacetates [2 and 3]: smaller amounts of vinyl derivatives [4, 5, 7, 8, 9] are also present and there is always some starting material [1]. Thus esterification and elimination reactions occur, the former predominating. The electronic, infra-red and nuclear magnetic resonance spectra of the various dicarboxylic acids have been recorded (11). The photophysical properties of excited states of some of the components have been determined by Professor Truscott and Dr. Land and their colleagues. Illustrative data are given in Table 2. In acetone solution the triplet lifetimes fall in a fairly narrow range, and the quenching of the triplet porphyrin by oxygen occurs efficiently and, indeed, approaches the diffusion controlled rate. The triplet quantum yields in acetone [for haematoporphyrin and haematoporphyrin disuccinate: ϕ_T = 0.92 and 0.95, respectively] are a little higher than those observed earlier for mesoporphyrin dimethyl ester [ϕ_T = 0.81] (13) or protoporphyrin dimethyl ester [ϕ_T = 0.80] (14) in benzene. However, the photophysical parameters of the metal-free porphyrins we have studied so far fall within a rather narrow range. Properties which determine localisation (solubility, partition coefficient, chemical reactivity) are likely to prove more important than photophysical parameters in determining biological activity *in vivo*.

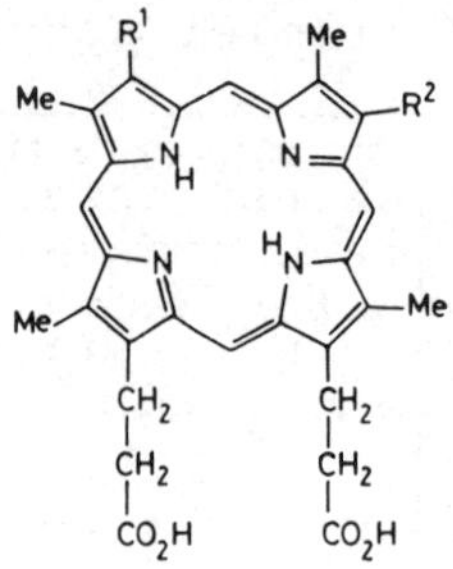

(**1**) $R^1 = R^2 = CH(OH)Me$
(**2**) $R^1 = CH(OH)Me$; $R^2 = CH(OAc)Me$
(**3**) $R^1 = CH(OAc)Me$; $R^2 = CH(OH)Me$
(**4**) $R^1 = CH(OH)Me$; $R^2 = CH{=}CH_2$
(**5**) $R^1 = CH{=}CH_2$; $R^2 = CH(OH)Me$
(**6**) $R^1 = R^2 = CH(OAc)Me$
(**7**) $R^1 = CH(OAc)Me$; $R^2 = CH{=}CH_2$
(**8**) $R^1 = CH{=}CH_2$; $R^2 = CH(OAc)Me$
(**9**) $R^1 = R^2 = CH{=}CH_2$

Figure 3. Structures of some components of HPD

We have devised an assay for biological activity *in vivo* to which all the fractions have been subjected (15,16). The assay is carried out with implants of the PC6 myeloma in mice. The porphyrin fractions are dissolved in a mixture of dimethyl sulphoxide and physiological buffered saline containing sodium bicarbonate and injected (6.2×10^{-5} moles/Kg) 3 hr after preparing the solution. The dimethyl sulphoxide is present to assist dissolution. After 24 hr the animals are prepared for irradiation, and are exposed for 10 minutes to heat-filtered white light (114 mW cm^{-2}) from a xenon arc. During this time the subcutaneous temperature does not rise more than 4°C. After 24 hours, Evans blue is administered in physiological saline, and the animals are killed two hours later. The tumours are fixed in formol-saline and sliced vertically, and the depth of necrosis is measured with a micrometer. In general six animals are employed in each test.

The results are shown in Table 3 (15,16). In this assay haematoporphyrin [1], hydroxyethyl vinyldeuteroporphyrin [4,5] and protoporphyrin [9] are found to have no detectable activity. The individual diastereoisomers of haematoporphyrin (α and β) are also inactive when tested separately. However, the acetoxy derivatives [2,3,6,7,8] all show activity. Although the activities vary somewhat, the monoacetate appearing to be the most active, we do not at this stage regard the test as making a clear distinction between the various acetates.

However, the situation turns out to be more complex

Table 3. *In vivo* biological activity of fractions of HPD eluted from HPLC column

Fraction numbers(11)	Identification*	Activity**
1-3, 9, 10	Polar materials + NaCl + interfraction cut	0 (6)
4-8	Haematoporphyrin [1]	0 (6)
11-13	Haematoporphyrin monoacetate [2,3]	1.02±0.29 (6)
14-16	Interfraction cut	0.22±0.33 (5)
17-19	Hydroxyethylvinyl deuteroporphyrin [4,5]	0 (6)
20-25	Haematoporphyrin diacetate [6]	0.72±0.46 (6)
26-33	Interfraction cut	0.47±0.16 (4)
34	Acetoxyethylvinyl deuteroporphyrin [7,8]	0.86±0.64 (5)
35-37	As 34, but additional minor components	0.51±0.38 (7)

* By analytical HPLC, and spectroscopic comparison with authentic compounds

** Depth of necrosis as a fraction of that obtained with HPD (number of animals in parenthesis).

than this. When haematoporphyrin diacetate [6] is injected in dimethyl sulphoxide alone it appears to be inactive in our test. When haematoporphyrin diacetate [6] is dissolved in dimethyl sulphoxide - PBS - sodium bicarbonate and injected within 1 minute it also shows no activity. However, the activity gradually develops with time and after 3 hours the full activity has developed. The monoacetate [2,3] shows similar behaviour.

We suppose that the treatment with mild aqueous base is activating these acetates. It will be recalled that HPD is conventionally administered after treatment with alkali (0.1

1. ESTER. Hydrolysed fairly readily. Elimination possible. Dimeric structure possible with four such bonds!

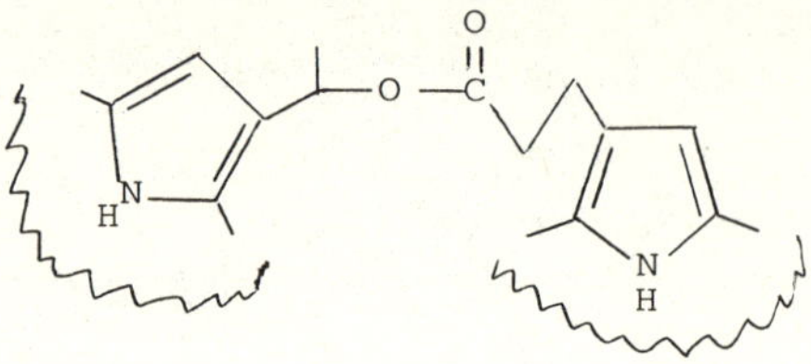

2. ETHER. More difficult to hydrolyse: elimination possible (to give vinyl and hydroxyethyl groups).

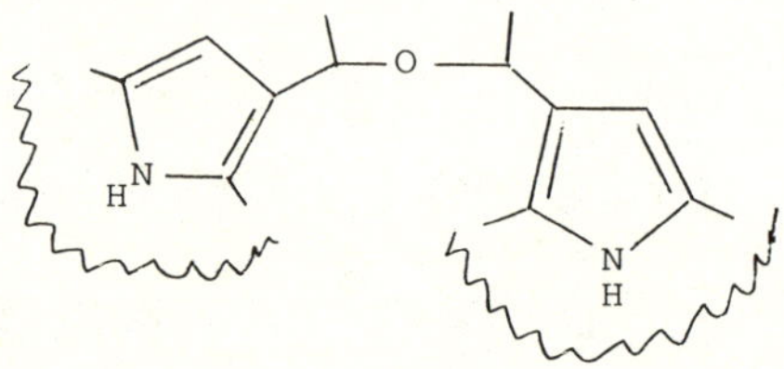

3. CARBON-CARBON BOND. Derived from electrophilic substitution of benzylic type carbonium ion at *meso* -position. Likely to be quite stable.

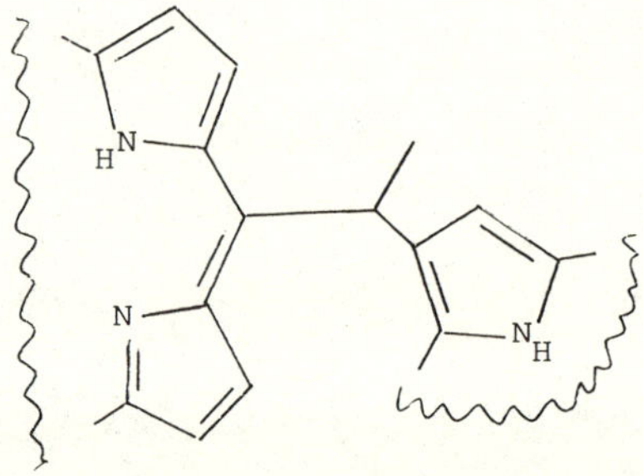

Figure 4. Possible modes of condensation of haematoporphyrin acetates to give intermolecular ester, ether, and carbon-carbon bonds.

M NaOH for 1 hour). This treatment rapidly destroys the acetate derivatives (10). Since we have now shown that the recognised products [1,4,5,9] of this reaction (10) are inactive in our assay, we conclude that other components present are the active species.

We presumed that these components of alkali-treated HPD were being retained by the HPLC column under the conditions we have used in the past (9,10). We have therefore fractionated alkali-treated HPD by reverse phase HPLC in the normal way, dividing the eluate arbitrarily into three fraction A, B, C. The spent column was then eluted with two powerful elutriants, fraction D: tetrahydrofuran:water: dimethyl sulphoxide and fraction E: 3% oxalic acid in ethanol. The various fractions were then subjected to the *in vivo* biological assay as described above. Since the HPD had already been "activated" by treatment with alkali, it was considered that further base treatment should be unnecessary.

The materials were therefore injected as solutions in dimethyl sulphoxide. Fraction A showed some activity: the main activity was found in fraction D, that is, the main material eluted from the "spent" column.

What is this material? Because it is retained on the column when polar and non-polar monomeric porphyrins of this series are eluted we suppose it to be a dimer or oligomer. It could be a physical aggregate - the phenomenon of aggregation of porphyrins in solution is well known - but because the reactive haematoporphyrin acetates are clearly the precursors of this material, we prefer to hypothesise that it is a covalently bonded dimer or oligomer. Structures containing intermolecular ester, ether, or carbon-carbon bonds (Fig. 4) can be advanced with varying degrees of support on chemical grounds, and we are now examining the constitution of this material in detail.

Finally, it seems appropriate to suggest that in future biological and clinical trials it would be preferable to use a single well defined substance, such as haematoporphyrin diacetate [6], rather than the complex and variable mixture represented by HPD.

ACKNOWLEDGEMENTS. This work was supported by the Medical Research Council. The photophysical work carried out by Professor T.G. Truscott, Dr. E.J. Land and their colleagues, is still in progress, and I am grateful to them for a selection of results ahead of full publication.

REFERENCES

1. W. Hausmann, Biochem. Z., 1911, 30, 276.

2. F. Meyer-Betz, Deutsch. Arch. Klin. Med., 1913, 112, 476.

3. H. Auler and G. Banzer, Z. Krebforsch., 1942, 53, 65; F.H.J. Figge, G.S. Weiland, and L.O. Manganiello, Proc. Soc. Exp. Biol. Med., 1948, 64, 640.

4. S. Schwartz, K. Absolon, and H. Vermund, Univ. Minnesota Med. Bull., 1955,27,7.

5. R.L. Lipson and E.J. Baldes, Arch. Dermatol., 1960, 82, 508; R.L. Lipson, E.J. Baldes, and A.M. Olsen, J. Natl. Cancer Inst., 1961, 26, 1.

6. I. Diamond, A.F. McDonagh, C.B. Wilson, S.G. Granelli, S. Nielsen, and R. Jaenicke, Lancet, 1972, ii, 1175.

7. J.F. Kelly, M.E. Snell, and M.C. Berenbaum, Brit. J. Cancer, 1975, 31, 237.

8. T.J. Dougherty, G.B. Grindey, R. Fiel, K.R. Weishaupt, and D.G. Boyle, J. Natl. Cancer Inst., 1975, 55, 115.

9. R. Bonnett, A.A. Charalambides, K. Jones, I.A. Magnus, and R.J. Ridge, Biochem. J., 1978, 173, 693.

10. R. Bonnett, R.J. Ridge, P.A. Scourides, and M.C. Berenbaum, J. Chem. Soc. Chem. Comm., 1980, 1198.

11. R. Bonnett, R.J. Ridge, P.A. Scourides, and M.C. Berenbaum, J. Chem. Soc. Perkin Trans I, 1981, 3135.

12. P. Clezy, T.T. Hai, R.W. Henderson, and L. van Thuc, Aust. J. Chem., 1980,33,545.

13. R. Bonnett, A.A. Charalambides, E.J. Land, R.S. Sinclair, D. Tait, and T.G. Truscott, J. Chem. Soc. Faraday I, 1980,76, 452.

14. R.S. Sinclair, D. Tait, and T.G. Truscott, J. Chem. Soc. Faraday I, 1980, 76, 417.

15. M.C. Berenbaum, R. Bonnett, R.J. Ridge, and P.A. Scourides, Abstracts, Tetrapyrrole Discussion Group, Glasgow, September 26th 1980.

16. M.C. Berenbaum, R. Bonnett, and P.A. Scourides, Br. J. Cancer, 1982, In Press.

17. R. Bonnett, C.R. Lambert, E.J. Land, P.A. Scourides, R.S. Sinclair, and T.G. Truscott, in preparation.

THE ANALYSIS OF HEMATOPORPHYRIN DERIVATIVE

Peter A. Cadby, Eugene Dimitriadis
Hamish G. Grant and A. David Ward

Department of Organic Chemistry
University of Adelaide,
Adelaide, South Australia, 5001

Ian J. Forbes

Department of Medicine
University of Adelaide
Queen Elizabeth Hospital
Woodville, South Australia 5011

INTRODUCTION

When work on photoradiation therapy (PRT) commenced in Adelaide, the stringency of Australian Federal drug regulations made it preferable for us to prepare HPD rather than import the product. It was therefore important that we have an acceptable assay for both HPD solid and HPD clinical[1], thereby enabling comparisons between our material and the product used elsewhere.

We initially examined the feasibility of separating porphyrin mixtures using conventional chromatography on silica or alumina. For this purpose, HPD solid was esterified with diazomethane. TLC on silica or alumina showed the presence of protoporphyrin (9, Fig. 1) and hematoporphyrin

---

[1]In this article, the term HPD solid refers to the product obtained from hematoporphyrin by acetylation with acetic:sulfuric acids; HPD clinical refers to the product of alkaline hydrolysis of HPD solid.

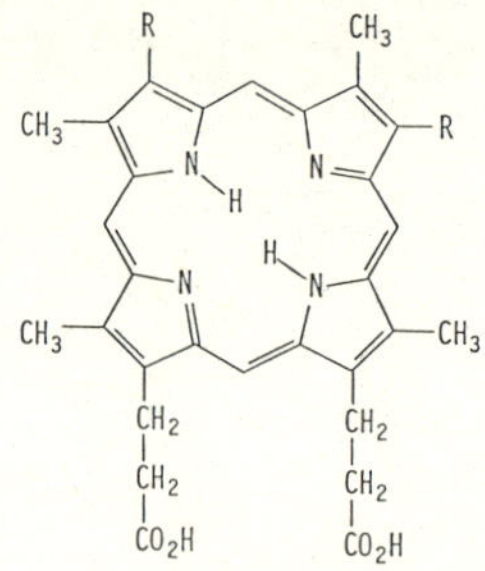

Figure 1. Porphyrins. [**1**] R = $CH_3CH(OH)$- (hematoporphyrin); [**2,3**] R = $CH_3CH(OH)$- & CH_2=CH-; [**4,5**] R = $CH_3CH(OAc)$- & $CH_3CH(OH)$-; [**6**] R = $CH_3CH(OAc)$- hematoporphyrin diacetate); [**7,8**] R = $CH_3CH(OAc)$ & CH_2=CH-; [**9**] R = CH_2=CH- (protoporphyrin).

[**1**] as the dimethyl esters, as well as other porphyrin products. Partial separation of these mixtures was achieved via repeated column chromatography or preparative TLC. NMR studies indicated the presence of porphyrins [**1-9**].

We also examined the separation of porphyrin free acids on silica. Both HPD solid and HPD clinical could be partially separated on silica plates, but considerable streaking was seen, especially with HPD clinical. Even pure compounds do not necessarily run well on silica. For example, protoporphyrin runs as a single spot with a $CHCl_3$ solvent, but shows considerable streaking, with material remaining at the origin, with methanol:benzene mixtures. The latter system has been suggested as a suitable solvent for porphyrin free acids on silica (1).

These separation methods were clearly not satisfactory for the quantitative analysis of HPD solid or HPD clinical, so we turned our attention toward the more sensitive and quantitative HPLC methodology.

<u>Separation of porphyrin esters by HPLC</u>

The porphyrin esters, prepared from HPD solid using diazomethane, could be resolved by HPLC using hexane: chloroform:ethanol as the solvent and Lichrosorb (10 μm) as adsorbent. The degree of separation could be altered by changing the ratios of the solvent components, particularly the % ethanol. For example, using hexane:chloroform:ethanol (70:27:3), the full chromatogram was obtained in 30 min, but the first two peaks (protoporphyrin dimethyl ester and hematoporphyrin diacetate dimethyl ester) were barely separated. When the ratio was 72:27:0.75, the retention time for HP dimethyl ester was 60 min (flow rate = 7 ml/min). These latter conditions were used for collection of individual components, where good separation rather than speed was the major consideration. In most cases, fractions were purified by further HPLC. Purified materials were analyzed by NMR spectroscopy and mass spectrometry; esters of porphyrins [1-9]. were thereby separated and identified. Our data are in agreement with results of Clezy et al. (2).

The results obtained from the separation and analysis of the esters do not necessarily provide quantitative or even qualitative information on the composition of the porphyrin acids in HPD solid and HPD clinical, since diazomethane may not quantitatively convert all porphyrins to their esters. Indeed, we always observed some acidic material remaining, even when excess diazomethane was used. Moreover, subsequent data showed that protoporphyrin was essentially absent from HPD solid, but was found in esterified samples of HPD solid, suggesting formation of protoporphyrin during some stage of the esterification or subsequent workup.

Separation and analysis of the porphyrin free acids

A published procedure (3) describes separation by HPLC of the components of HPD solid on a reverse-phase column using a solvent mixture of methanol and water containing an ion-pairing reagent. Our initial attempts to reproduce this separation were only partly successful, and indicated considerable differences between our optimum conditions and those reported. We therefore undertook an investigation on effects of changing individual parameters to ascertain determinants of the separation technique.

After experimentation, it became apparent that different packing materials showed a wide variation in separation capability. Best results were achieved with commercial columns: the Waters Radial Pak A (reverse-phase C-18) column containing 5 μm adsorbant, and the Waters μ-Bondapak C-18 (10 μm) columns. We attribute this difference to the extent to which the silica in each column has been silanized. With the μ-Bondapak column, any silica remaining unreacted after treatment with the C-18 silylating reagent is 'end capped' using trimethylsilyl chloride. Radial Pak columns do not receive this treatment. Hence the type of silica, mesh size and silylation procedure are factors which can produce column materials whose performance can vary considerably, yet provide good resolution of porphyrin acids.

Initially, two solvent systems were found satisfactory: methanol-water (approx. 8:2) and acetonitrile-water (approx 6:4). Both solvent systems give best results at low pH values (with phosphoric or acetic acid added) although lower pH values lengthened retention times. For example, proto-

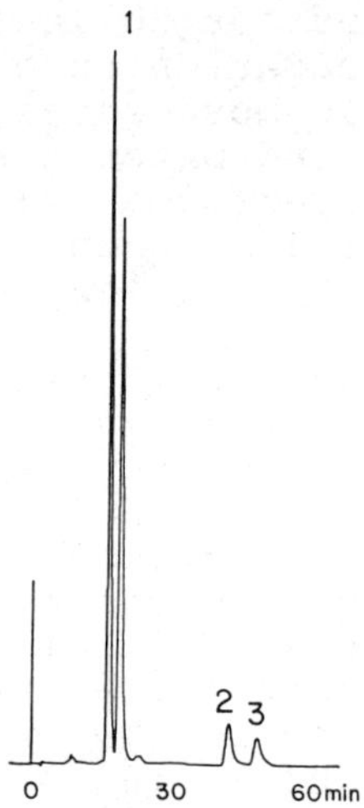

Figure 2. Chromatogram of hematoporphyrin diHCl in dimethyl sulfoxide solution. Peak numbers correspond to structural numbers in the text and in Fig. 1. Column: Waters Radial-Pak C-18 (5 μm). Solvent: 80% MeOH, 20% aqueous IPR at pH 2.0. Flow rate = 1.5 ml/min.

porphyrin has a retention time of 60 min at pH 2, but of 15 min at pH 5 (Radial Pak column) with other conditions identical. A decreased retention time generally decreased resolution. At higher pH values (4-7), the chromatogram often consisted of some sharp but tailing peaks superimposed on several broad, ill-defined peaks. This behavior, observed with different columns and solvents, is tentatively attributed to aggregation which readily occurs with porphyrin acids. It is known (4) that the porphyrin monomer is substantially favored at lower pH values, and we suggest that the broad unresolved peaks observed at higher pH are caused by dimers and higher aggregates.

For either of the solvent systems described above, separation was considerably improved by addition of an ion-pairing agent. While triethylammonium phosphate (at the appropriate pH) was satisfactory, better results were obtained with tetra-n-butylammonium phosphate which was used for most separations.

Ion-pairing reagents have been used extensively in separations of polar, water-soluble materials on reverse-phase columns. The role of the reagent is poorly understood (5). The substantial improvement obtained with the Radial Pak column suggests that ion-pairing reagents might function

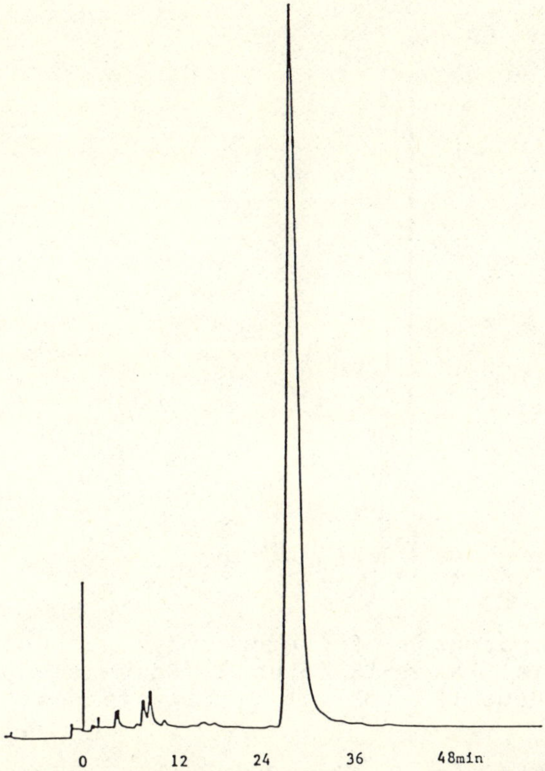

Figure 3. Chromatogram of HPD solid. Column: Waters Radial-Pak C-18 (5 μm). Solvents: [A] acetonitrile, aqueous IPR (pH 2.5), MeOH (3:2:1). [B] methanol, aqueous IPR pH 2.5 (9:1). Flow rate = 2 ml/min.

to some extent as alternatives to 'end capping' by absorption onto unprotected OH groups on the silica.

The methanol:water system consistently gave two peaks of approx. equal response for hematoporphyrin on a variety of columns. This complete resolution of two very close peaks (Fig. 2) was not achieved with acetonitrile:water which afforded only partial separation. The two peaks were observed with HPD (solid or clinical) and in commercial hematoporphyrin samples from two different sources, and with hematoporphyrin dimethyl ester analysed using the methanol: water system. The two components of hematoporphyrin and of the ester have been separated and collected using semi-preparative conditions and produced single peaks upon re-injection. This result is not compatible with the expected behaviour of two equilibrating isomers. A fraction containing only the two peaks associated with the hematoporphyrin dimethyl ester was collected, hydrolyzed (pH 12, 10 hr at

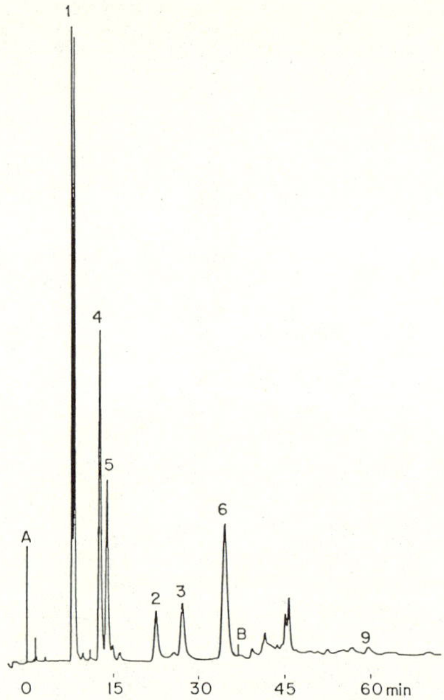

Figure 4. Chromatogram of protoporphyrin. Column: Waters Radial-Pak C-18 (5 μm). Solvent: 90% MeOH, 10% aqueous IPR (pH 2.5). Flow rate = 2 ml/min.

room temperature), neutralized and the product analysed by HPLC. Two peaks corresponding to the HPLC behaviour of hematoporphyrin were obtained. Mass spectral analysis of each separated component from the ester gave a molecular ion at m/e 626 corresponding to that of hematoporphyrin dimethyl ester, and showed the expected fragmentation pattern for a compound with two hydroxyethyl side chains. We therefore suggest that the two peaks correspond to diastereoisomers of hematoporphyrin due to the presence of the two chiral carbon atoms in the hydroxyethyl side chains.

The nine components, [1-9], Fig. 1, present in HPD solid differ only in the nature of the side chains (R_1 and R_2) and it is convenient to refer to them by indicating the nature of these side chains. Thus, [2] and [3] containing a 2-hydroxyethyl and a vinyl side chain are designated hydroxy-vinyl derivatives; [4] and [5] are the hydroxy-acetates, and [7] and [8] are the vinyl-acetates. Using this system, hematoporphyrin [1] is a dihydroxy compound, protoporphyrin [9] is divinyl, and the diacetate derivative [6] of hematoporphyrin becomes the diacetate. The advantage of these designations is that it is then possible to discuss the isomeric pairs, e.g., [2] and [3] without necessarily

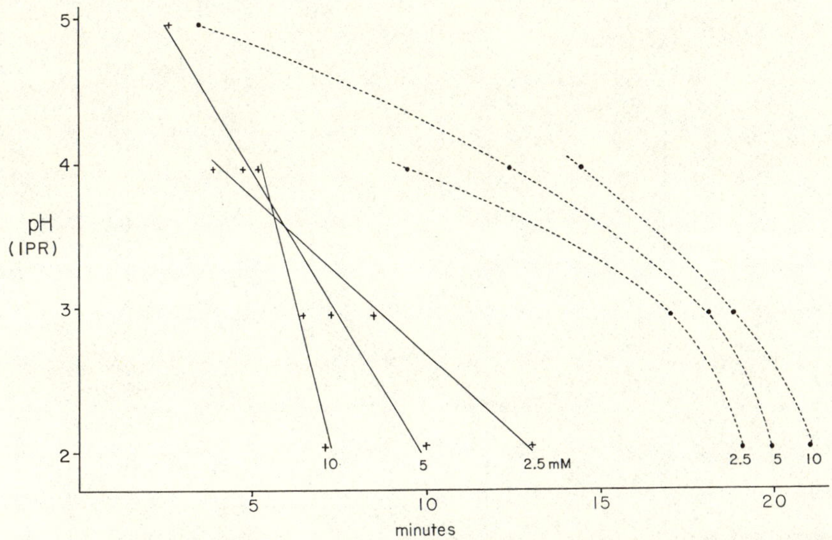

Figure 5. Plot showing the effect of aqueous IPR pH and concentration on retention time of hematoporphyrin (solid line) and protoporphyrin (dashed line, multiply time scale X 3. Column: Waters Radial Pak C-18 (5 μm). Solvent = 80% MeOH, 20% IPR. Flow rate = 2 ml/min.

having to indicate their exact structures.

The peaks in the chromatograms of HPD solid and HPD clinical can readily be identified. The two hydroxy-vinyl isomers [2] and [3] are always present to some extent in commercial hematoporphyrin enabling their retention volume to be determined (Fig. 2). Acetylation of hematoporphyrin using acetyl chloride:pyridine yields the diacetate [6] as the major product. The diacetate can be separated from the other products by careful chromatography on silica. Minor products from the acetylation are the hydroxy acetates [4] and [5] and the vinyl acetates [7] and [8]. These pairs can be separated from other components by HPLC (Fig. 3), esterified and identified via spectral methods (6). We have not attempted to ascertain which peak of each particular pair of isomers belongs to which structure. Protoporphyrin can readily be prepared from hematoporphyrin. The analysis of this product by HPLC (Fig. 4) clearly indicates its high purity.

Notwithstanding the highly acceptable performance of the binary solvent systems (methanol or acetonitrile +

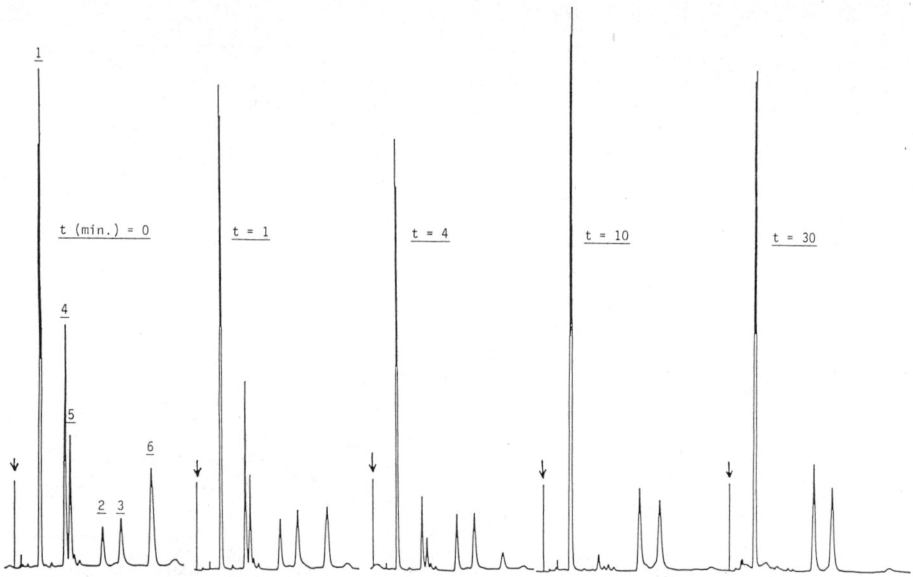

Figure 6. Change of composition of HPD solid with time upon exposure to 0.1 N NaOH. Arrows indicate injection points. Column: Waters Radial-Pak C-18 (5 μm). Solvent [A]. Flow rate = 2 ml/min.

aqueous IPR[2]) for analyzing HPD clinical, these systems were less than ideal for analysis of more complex mixtures, e.g., HPD solid. For example, with the methanol-based system, the diacetate peak [6] had a retention time identical to that of one of the hydroxy-vinyl isomers, and the isomeric hydroxy-acetates were not resolved. The acetonitrile-based system (60% CH_3CN + 40% aqueous IPR, pH 2.5) provided greatly improved separation of polar components at the tolerable expense of complete loss of resolution of hematoporphyrin diastereoisomers.

Aqueous acetonitrile containing IPR gave no resolution of hematoporphyrin peaks on the analytical μ-Bondapak column. Some resolution of this peak had been previously obtained using the aqueous acetonitrile:acetic acid system on this column, indicating that strong hydrogen-bonding agents are required for selective HPLC separation of the hematoporphyrin diastereoisomers. Addition of 10-20% methanol to the aqueous acetonitrile solvent afforded not only partial resolution of the two hematoporphyrin components, but also improved separation of the other components in this

[2]IPR=ion-pairing reagent.

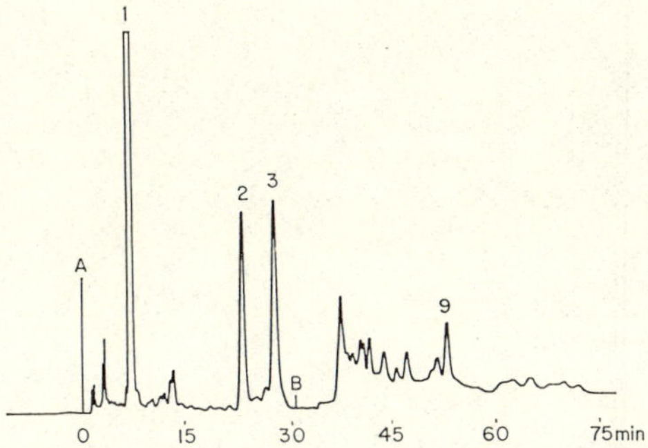

Figure 7. Chromatogram of HPD clinical. Conditions as in Fig. 3.

region, notably the hydroxy-acetates. In this system, the peak attributed to the diacetate [6] had a longer retention time than did the hydroxy-vinyl isomers (Fig. 3).

Despite the considerably improved resolution and peak shape afforded by the ternary system in the 'polar region', the protoporphyrin peak showed considerable tailing and was virtually undetectable in <5-8% concentration. This problem was overcome by utilizing a dual solvent system whereby the aqueous acetonitrile:methanol: IPR (pH 2.5) solvent was changed to 90% aqueous methanol:IPR (pH 2.5) after elution of the second hydroxy-vinyl (or diacetate) peak. This procedure improved peak shape, provided a convenient retention volume for protoporphyrin and revealed, via a decreased resultant peak width, a number of previously-obscured minor components. The majority of the latter were found in the region between the diacetate and protoporphyrin, but several minor peaks were eluted after protoporphyrin which is usually considered to be the least polar compound of interest.

It may be noted that acetonitrile, possessing a relatively low hydrogen-bonding potential, enhanced resolution of the more polar components as a consequence of minimized solvent-solute interactions. Methanol appears to provide better resolution and peak shape for the less polar components eluting after the diacetate. The dual solvent approach appears to offer an acceptable compromise for convenient HPD analysis if gradient elution is not available. Fig. 5 illustrates the effect of changing concentration and pH of the IPR on porphyrin retention times. The retention time of both hematoporphyrin and protoporphyrin decreased substantially with increasing pH. Changes in the concentration of the IPR produce opposite effects on these two porphyrins. For hematoporphyrin (a polar compound), retention time

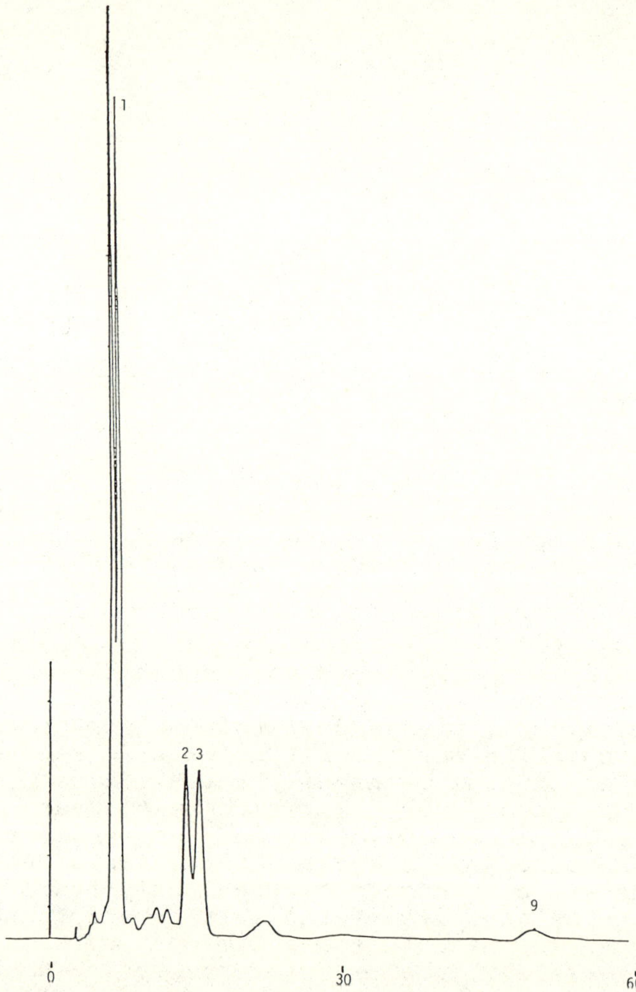

Figure 8. Chromatogram of HPD clinical. Column: Waters μ-Bondapak (10 μm). Solvent: MeOH, water, acetic acid (80:20:4). Flow rate = 1 ml/min.

decreases as the IPR concentrating increases (at low pH). With protoporphyrin (a non-polar compound), an increase in IPR concentration causes an increased retention time. This situation reverses with hematoporphyrin at high pH, suggesting that the IPR may interact with porphyrins as well as the column surface.

HPD clinical is prepared from HPD solid by treatment with dilute NaOH; there has been some discussion on the time necessary to complete this process. In Fig. 6, we show changes with time in composition of HPD solid upon treatment with base. Disappearance of diacetate **[6]**, a major component of HPD solid, was essentially complete after 5 min. The rate of hydrolysis of the hydroxy acetates was slower, presumably since these are formed during diacetate hydrolysis, but monoacetates are not detectable after 20 min at which time the chromatogram resembles that of HPD clinical (Fig. 7) for which a hydrolysis time of 60 min is considered necessary. Additional time of exposure to base may be necessary for physical changes involving the porphyrins to occur in order to maximize the effect of HPD clinical.

Protoporphyrin is barely detectable in HPD solid, but is present, although to a minor (*ca.* 3%) extent in HPD clinical (Fig. 7). Alkaline treatment of HPD solid must cause some elimination of acetate groups as well as hydrolysis. Since the hydroxy-vinyl isomers **[2]** and **[3]** are major components of HPD clinical, it is possible that the latter product represents a convenient means of obtaining, in aqueous solution, a significant concentration of porphyrins with vinyl side chains.

Fig. 7 shows the complete trace for a sample of HPD clinical and displays the base-line shift obtained with the solvent change. Fig. 8 illustrates the effect of changing the column material: compared with the Radial-Pak column, the μ-Bondapak column separates the hydroxy-vinyl isomers much less adequately, and poorly resolves the protoporphyrin peak.

Integration of the chromatogram for HPD clinical (Fig. 4) shows presence of hematoporphyrin (45-50%), hydroxy-vinyls (20-25%) and protoporphyrin (3-5%). Minor peaks eluting near hematoporphyrin make up 5-10% of the total; minor peaks eluted with the more polar solvent but excluding protoporphyrin account for 10-15%. This last figure is less accurate than the others because of the difficulty of correcting for base-line shift upon changing solvents.

HPD solid is much more variable in composition because of changes which occur on storage, especially if the material is not adequately washed and dried. The diacetate **[6]** is the major component, >50% of freshly-prepared material. Depending on storage conditions, this value may drop considerably. In spite of these changes, the analysis of HPD clinical prepared from HPD solid does not vary substantially, suggesting that changes occurring during storage of HPD solid are similar to those caused by the alkali treatment. The major difference is that the relative amount of the hydroxy-vinyls is increased in HPD clinical when it is prepared from fresh HPD solid.

EXPERIMENTAL

All solvents were distilled, degassed and filtered through a 0.45 µm Millipore filter before use. HPLC was performed using a Waters 600A solvent delivery system and U6K injector. The detector normally used was a Waters model 440 UV absorbance type operating at 405 nm. Columns used were Waters µ-Bondapak (10 µm) and Waters Radial-Pak C-18 5 mm and 8 mm i.d. (5 and 10 µm).

Hematoporphyrin was obtained as the diHCl from Roussel. Protoporphyrin was prepared from hematoporphyrin by brief heating in dimethylformamide. Hematoporphyrin diacetate was prepared by treating hematoporphyrin with acetic anhydride and pyridine and was approx 80% pure. Hematoporphyrin dimethyl ester, prepared via diazomethane, was separated by HPLC into two peaks, resolved to the base line, using conditions described in Fig. 2. Each peak was collected, the solvent removed, and the residue analyzed by mass spectrometry (AEI MS-30).

The ion-pairing reagent was tetra-n-butylammonium phosphate (Unichrom). A 2.5 mM solution in distilled water was prepared and the pH adjusted with phosphoric acid.

ACKNOWLEDGMENTS

We thank Mr. M. Rechsteiner (Waters Associates) for his advice and Mr. L. Paltridge for technical assistance. The work was supported by grants from the University of Adelaide, the Anti-Cancer Foundation of the Universities of South Australia and the National Health and Medical Research Council. We thank Miss E. Laubmann for a generous donation to the University of Adelaide Cancer Phototherapy Fund.

REFERENCES

1. D Kessel, Cancer Res., 41,1318 (1981).

2. PS Clezy, TT Hai, RW Henderson and LV Thuc, Aust. J. Chem, 33, 585 (1980).

3. R Bonnett, AA Charalambides, K Jones, IA Magnus and RJ Ridge, Biochem. J., 173, 693 (1978). R. Bonnett, RJ Ridge, PA Scourides and MC Berenbaum., J. Chem. Soc. Chem. Commun. I., 1198 (1980) and J. Chem. Soc. Perkin I, 3135, (1981).

4. WI White, in "The Porphyrins, Physical Properties, Part C". Ed. D. Dolphin. Academic Press, NY 1978, Chapter 7.

5. BA Bidlingmeyer, SN Deming, WP Price, B Sachole and M Petrusek, J. Chromat., 186, 419 (1979).

6. PS Clezy, CJR Fookes and TT Hai, Aust. J. Chem., 31, 365 (1978).

POTENTIAL USE OF RADIOLABELLED PORPHYRINS FOR TUMOR SCANNING

R.A. Thaller and D.M. Lyster

Division of Nuclear Medicine
Vancouver General Hospital and
Department of Pharmacy
University of British Columbia
Vancouver, B.C., Canada

D. Dolphin

Department of Chemistry
University of British Columbia
Vancouver, B.C., Canada

INTRODUCTION

The selective biodistribution of metal-free porphyrins was first noted by Policard in 1924 based on the observation of a "spontaneous appearance" of fluorescence in tumors. It was then shown (Auler and Banzer, 1942; Figge, 1945) that parenteral administration of several naturally occurring porphyrins including hematoporphyrin (HP) [1], protoporphyrin [2], mesoporphyrin [3] and the zinc complex of [1] resulted in an accumulation of these porphyrins in subcutaneous sarcomas causing the tumor to fluoresce an orange-red color under ultra-violet irradiation. In addition, other nontumor sites including injury sites, placenta, developing embryos and lymph nodes (Manganiello <u>et al.</u>, 1951) accumulated some porphyrin. Another more detailed study showed that metalloporphyrins were taken up by tumors (Peck <u>et al.</u> 1953). The first clinical study using HP [1] by Peck <u>et al.</u> in 1953 showed no tumor uptake. Injected doses were kept low (less than 120 mg/patient) due to the reported toxicity of the drug (Meyer-Betz, 1913). This toxicity, however, was caused by phenol in the product not the agent

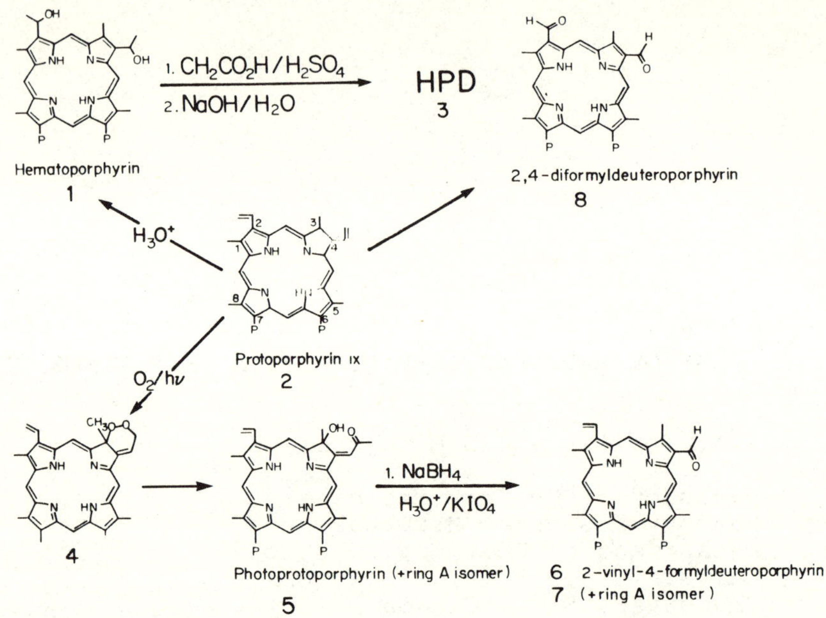

itself. Definite tumor-uptake was reported using phenol-free HP [1] at doses of 500-1000 mg/patient with no toxic effects (Rassmussen-Taxdal et al., 1955). When the porphyrin was infused over a period of 3-12 hr, squamous cell carcinoma of the tongue and penis; adeno-carcinoma (sigmoid colon, rectum, ascending colon); olfactory-grove meningioma and carcinoma of the breast all showed porphyrin uptake.

Adenocarcinoma of the prostrate, ependynoma of the cervical cord, and fibrotic abscesses of the breast did not show porphyrin localization. It was further reported (Rassmussen-Taxdal et al., 1955) that porphyrin localization and fluorescence was useful in tumor detection and visualization during surgery. Commercial HP is usually contaminated with other porphyrins (Schwartz et al., 1965). A HP derivative (HPD; 3) was prepared by dissolving the crude mixture in glacial acetic acid: concentrated sulfuric acid (1:15), filtering and precipitating the derivative by neutralization with 15 to 20 volumes of 3% sodium acetate. The derivative was a better tumor localizing agent than the crude mixture with lower toxicity limits and patient doses. There was also minimal accumulation in lymphatic tissue, benign fibroadenomas and fresh wounds probably due to serum concentration in the area. The derivative was not taken up by granulomatous lesions nor did it pass through the placental barrier to the fetus. However, the uterus and membrane did show fluorescence (Lipson et al., 1961).

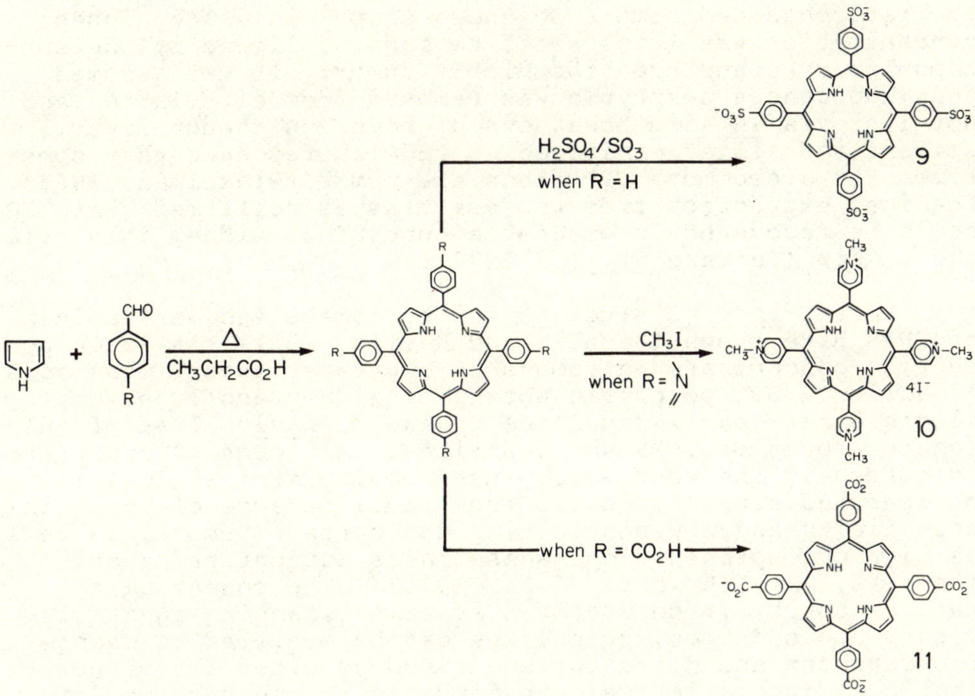

It is clear that HPD is a complex mixture of porphyrins. Murine leukemia L1210 show preferential *in vitro* uptake of the more hydrophobic components of HPD (Kessel, 1981). This system exhibits good photosensitization but porphyrin was readily removed from cells by washing. In contrast, the more hydrophilic components of HPD only gradually accumulated in the tumors. Nevertheless, these complexes appeared to be responsible for the preferential affinity for neoplastic cells. The unnatural porphyrin, tetraphenylporphine sulfonate (TPPS; 9) was studied by Winkelman (1962). TPPS was found to be more highly concentrated in Walker carcinosarcoma than any other tissue of the rat; and was better,concentrated, by an order of magnitude, than HP. At autopsy there was brilliant red fluorescence in the tumor, bright red fluorescence distributed irregularly in the lung, faint red fluorescence in the lymph nodes and pancreas. No fluorescence was apparent in the liver, kidney or spleen. Quantitative analysis using fluorescence and absorption spectrophotometry showed considerable amounts in liver, kidney, and spleen, followed by lung. However, small amounts were present in all tissues assayed (total 18); with skin and muscle always having the least amount. It was also shown that the Walker carcinosarcoma of the rat contained some endogenous porphyrin. The highest concentration was in necrotic regions, followed by the subcapsular area and then the viable tumor. It was assumed that endogenous porphyrin was removed from circulation and not lost via *in situ* breakdown of heme. Although liver, spleen, and kidney exhibited no red fluorescence they sometimes contained more TPPS than the tumor (Winkelman, 1963). Chemical extraction from various tissues confirmed that TPPS could be recovered in greater amounts from kidney than from the tumors (Carrano *et al.*, 1977).

Recent *in vivo* studies with radiolabelled derivatives of TPPS have shown (Zanelli and Kaelin, 1981) that such porphyrins concentrate in tumors to a greater extent than does ^{67}Gallium. But porphyrin uptake by kidney and lungs was also greater than was gallium uptake. *In vivo* loss of sulfonate groups of TPPS was described. Although ^{67}Ga-citrate is currently the most widely used radiopharmaceutical in nuclear medicine, it is far from ideal because of both nonspecificity and low sensitivity for certain tumors, as well as high gut uptake which necessitates patient preparation (Andrews, G.A., Edwards, C.L., 1975). The potential of various porphyrin derivatives as tumor scanning agents is clear. In addition, porphyrins can be modified to change biotransport and distribution. Studies cited above show that changing peripheral substituents is one such modification. A second modification is the incorporation of a metal into the porphyrin.

Initial studies with metalloporphyrin derivatives suggested that the incorporation of a metal did not destroy tumor-seeking properties (Figge *et al.*, 1948). Thus ^{64}Cu-protoporphyrin was shown to concentrate in mouse tumors despite its lack of good tumor to liver, blood, or muscle ratios (Bases, 1954). Human studies showed no tumor uptake of this agent. It was also concluded (Bases *et al.*, 1967) that ^{57}Co-hematoporphyrin might be useful for detection of tumors not located near the liver, spleen or RE systems. More recent experiments with ^{57}Co, ^{99}Tc and ^{109}Pd derivatives of TPPS and HP indicated no tumor concentration. Since the majority of metalloporphyrins do not fluoresce, their detection and quantitation is difficult at low concentrations. This phenomenon could account for conflicting observations on tumor-localization by metalloporphyrins. From these observations and those described above it is apparent that while many porphyrin derivatives preferentially accumulate in tumors, factors affecting subsequent biotransport and distribution are unknown. Effects of any subsequent porphyrin metabolism on localization are also uncertain.

We report here on the synthesis of a variety of metalloporphyrins coordinated to γ-emitting nuclides of short half life. Derivatives of protoporphyrin and of totally synthetic porphyrins have been examined for their *in vitro* and *in vivo* tumor localizing properties.

MATERIALS AND METHODS

Porphyrins

Hematoporphyrin dihydrochloride was obtained from ICN Pharmaceuticals, Cleveland, OH or Sigma Chemical Co., St. Louis, MO. Meso-tetra(4-carboxyphenyl)porphyrin [11]), meso tetra(4-N-methylpyridyl)porphyrin tetraiodide [10], and tetra-Na-meso-tetra(4-sulfatophenyl)porphyrin [9] were obtained from Strem Chemicals Inc., Newburyport, MA.

Chromatography

Silica gel for column chromatography (Woelm activity I, 70-150 mesh, cat. no. 402747) was obtained from ICN Pharmaceuticals, Cleveland, OH. Polyamide for thin layer and column chromatography was obtained from Brinkmann Instruments, Westbury, NY. Dowex AG 1-X8 anion exchange resin and basic alumina AG10 were obtained from Bio-Rad Laboratories, Richmond, CA.

Radionuclides

^{59}Fe-ferric chloride was obtained from Amersham Corp., Arlington Heights, IL. All other chemical were of reagent grade or better. Dichloromethane and *t*-butanol were dried by distillation from calcium hydride.

A. Porphyrins

Protoporphyrin, photoprotoporphyrin [5] (isomer 1: 1-hydroxy-2-desvinyl-2-formylethylideneprotoporphyrin; ring B reacted: isomer 2: 3-hydroxy-4-desvinyl-4-formylethylidene-protoporphyrin DTBE; ring A reacted), 2-formyl-4-vinyldeuteroporphyrin (Spirographis porphyrin [7], 2-vinyl-4-formyl-deuteroporphyrin [6], 2,4-diformyldeuteroporphyrin [8] and their DTBE derivatives were prepared by standard procedures (DiNello and Chang, 1978). In each case the ditertiary butyl esters were initially prepared. These complexes are readily soluble in organic solvents which allows for their chromatographic purification. Once pure, the ester groups are readily cleaved by acid to give the *pure* porphyrin dicarboxylic acid. This is a far superior route since the purification of the water soluble porphyrin dicarboxylic acids is both tedious and difficult. The following example shows how the deesterification step is performed.

Diformyldeuteroporphyrin free acid [8]

Ten mg of diformyl deuteroporphyrin DTBE were dissolved in 5 ml of dry dichloromethane. Anhydrous HCl was bubbled through the solution for about 6-10 hr. The progress of the reaction was monitored by TLC (dichloromethane ether 20:1). After completion of the reaction the solution was evaporated to dryness. The free acid residue was dissolved in 3 ml of pyridine. Addition of 5 ml of hot acetic acid resulted in immediate crystallization of the diformyl deuteroporphyrin. Purity was checked by TLC analysis and absorption spectrophotometry.

B. Metalloporphyrins

^{59}Fe-Hematohemin

The ferrous sulfate or chloride method was used (Falk, 1964). HP (10 mg) was dissolved in a solution containing 0.5 ml of pyridine, 10 ml of glacial acetic acid and 10 mg of ascorbic acid in a pear-shaped or long necked flask. Nitrogen was bubbled through the solution which was heated in an oil bath to 80°. The ^{59}Fe (100 μCi) as ferric chloride or citrate was added and the temperature raised to 90°. After 5 min., 1.0 ml of saturated ferrous chloride or sulfate was added. After an additional 5 min. at 90° the nitrogen was turned off and the solution allowed to cool to room temperature. Air was then passed through the solution for 30 seconds. This solution was then added to 10 ml of distilled water and 20 ml of chloroform. After extraction the organic layer was washed with 0.2 M HCl and then distilled water several times. The hemin was extracted from the organic layer into 5 ml of 1.0 N NaOH. The solution was adjusted to pH 7-8 with about 5 ml of 1.0 N HCl. Progress of the reaction was checked by paper chromatography with ethanol, and by absorption spectrophotometry. Radioactivity was also measured in each fraction to determine reaction yields. The final product was checked by absorption spectrophotometry, paper chromatography and TLC using n-butanol-water-acetic acid (50:1.5:1.4) (protohemin Rf=0.47, hematohemin Rf=0.34).

^{59}Fe-Protohemin

This procedure was similar to hematohemin except that protoporphyrin replaced HP. The product was crystallized from chloroform:methanol. The same quality control procedures were carried out on protohemin as on hematohemin.

^{59}Fe-Photoprotohemin

Photoprotoporphyrin (75 mg) was dissolved in 10 ml of

DMF. The solution was heated to 100° under N_2. ^{59}Fe in the form of ferric chloride or citrate (100 Ci) was added. After 15 min. of constant stirring, 20 gm of ferric chloride was added. After an additional 30 min. of heating and stirring the solution was cooled to room temperature and DMF removed by evaporation. The hemin residue was dissolved in minimal pyridine and an equal amount of 90-100% formic acid added. The hemin was precipitated by addition of ether. The precipitate was filtered and washed with ether, air dried, washed with water, air dried, washed with ether again, and then dried to constant weight.

The crude protohemin was purified by both thick layer and thin layer chromatography. Crude material (50 mg) was dissolved in minimal pyridine containing 5% of 90-100% formic acid, and applied to a 30 cm x 30 cm thick silica gel plate. The plate was developed in BMF (110:30:1). The green photoprotohemin band (middle band) was scraped off, pulverized, and the hemin diluted with methanol. The solution was reduced to 2 ml and 25 ml of petroleum ether were added. The hemin was precipitated, filtered, and washed as for the crude photoprotohemin as above. It was again purified on silica TLC plate as above except after chromatography, the methanol solution was taken to dryness. The purified photoprotohemin was dissolved in 5 ml of 1.0 N NaOH and the pH adjusted to pH 7-8 with 1.0 N HCl. The progress of the reaction and purity of the final product were monitored by paper chromatography with ethanol, absorption spectrophotometry, TLC using BMF (110:30:1) and polyamide TLC using BMF and methanol-acetic acid (100:2).

<u>^{59}Fe-2-formyl-4-vinyl deuterohemin</u>

Thirty mg of 2-formyl-4-vinyl deuteroporphyrin were added to 10 ml of DMF at reflux under nitrogen. ^{59}Fe as ferric chloride or citrate was added followed by 15 mg of ferric chloride 15 min. later. The progress of the reaction was followed by paper chromatography and absorption spectrophotometry. When the reaction was complete (45-47 min) the mixture was taken to dryness by rotary evaporation. The crude hemin was precipitated by ether, washed and purified by thick layer silica gel chromatography as described for ^{59}Fe-photoprotohemin.

The purified hemin was extracted from the silica gel by BMF (110:30:1). The solvent mixture was slowly removed by rotary evaporation until the product precipitated. The precipitate was washed with benzene and air dried. The purified hemin was crystallized by dissolving it in 1.5 ml of pyridine and 3 ml of chloroform (35:100). The filtrate and washings were combined, heated to boiling and 25 ml of boiling acetic acid were added; 0.35 ml of concentrated HCl were added and the mixture heated to boiling again. The solution was allowed to stand for 24-48 hr to complete crystallization. The hemin was recovered by suction filtration. The 2-formyl-4-vinyl deuterohemin was dissolved in 5 ml of 1.0N NaOH and the pH adjusted to pH 7-8 with about 5 ml of 1.0 N HCl. The purity was checked by absorption spectrophotometry and silica gel and polyamide TLC.

^{59}Fe-2-vinyl-4-formyl deuterohemin

This procedure was similar to 2-formyl-4-vinyl deuterohemin except that 2-vinyl-4-formyl deuteroporphyrin was used instead of 2-formyl-4-vinyl deuteroporphyrin.

^{59}Fe-2,4-diformyl deuterohemin

Thirty mg of diformyl deuteroporphyrin were metalated using the ferric chloride method (refluxing in DMF) as previously described. The resulting residue was dissolved in minimal methanol. The addition of water caused the crude product to precipitate. The precipitate was air dried, washed with ether, and air dried again. The crude hemin was dissolved in 5 ml of pyridine containing 10% acetic acid. This solution was streaked out on a thick layer silica gel plate and dried. The preloaded silica gel was scraped off, pulverized, and suspended in BMF (110:30:1). The slurry was poured onto a 25 g polyamide column equilibrated with the same solvent system. The column was eluted with BMF (110:30:1) until the minor band was eluted. The hemin was eluted off the column with BMF (110:70:1). The band was taken to dryness and crystallized as described for 2-formyl-4-vinyl deuterohemin.

^{59}Fe-meso-tetra(4-carboxyphenyl) hemin (TCP)

Ten mg of meso-tetra(4-carboxyphenyl) porphine were dissolved in 10 ml of glacial acetic acid containing 10 mg ascorbic acid and 10 mg of sodium acetate. The solution was heated to reflux under nitrogen. ^{59}Fe in the form of ferric chloride was added and refluxed for 30 min. Then, 1 ml of saturated ferrous chloride solution was added and refluxed for an additional 30 min.. The nitrogen was turned off and the solution cooled to room temperature. Air was passed through the solution for 1-2 min and the solution was adjusted to pH 7-8 with sodium bicarbonate. Incorporation of label was measured using paper chromatography.

^{59}Fe-meso-tetra(4-N-methylpyridyl) hemin tetraiodide

This procedure was similar to ^{59} Fe-meso-tetra(4-carboxyphenyl) hemin preparation with meso-tetra(4-N-methylpyridyl) porphine tetraiodide replacing meso-tetra(4-carboxyphenyl) porphine.

^{59}Fe-tetra-Na-meso-tetra(4-sulfonatophenyl) hemin (TPPS)

This procedure was similar to ^{59}Fe-meso-tetra(4-carboxyphenyl) hemin preparation with tetra-Na-meso-tetra(4-sulfonatophenyl)porphine (12 hydrate) replacing meso-tetra(4-carboxyphenyl) porphine.

Nickel spallation product labelled meso-tetra(4-N-methylpyridyl)porphyrin tetraiodide

During the bombardment of nickel with high energy protons (500 MeV) at the TRIUMF cyclotron many radionuclides are produced by the spallation process. ^{59}Fe was readily separated and coordinated to porphyrins (see below) in addition, various spallation products including ^{57}Co, ^{51}Cr and ^{54}Mn were studied. The nickel target was dissolved in HCl and evaporated in the presence of nitric acid. The residue was used to metallate meso-tetra(4-N-methylpyridyl) porphyrin tetraiodide in refluxing DMF. After labelling, the DMF solution was evaporated to dryness and the porphyrins resuspended in water.

^{52}Fe-protohemin

This procedure was similar to ^{59}Fe-protohemin except that ^{52} ferric chloride was used instead of ^{59}Fe ferric chloride.

^{52}Fe-meso-tetra(4-carboxyphenyl) hemin

This procedure was similar to ^{59}Fe-meso-tetra(4-carboxyphenyl) hemin except that ^{52} Fe ferric chloride was used instead of ^{59}Fe ferric chloride. A simpler procedure was also used in which the solution was heated to boiling not under nitrogen and the ascorbic acid and ferrous chloride were not added.

^{52}Fe-meso-tetra(4-N-methylpyridyl) hemin tetraiodide

This procedure was similar to ^{52}Fe-meso-tetra(4-carboxy-phenyl) hemin preparation except meso-tetra(4-N-methylpyridyl) porphine tetraiodide replaced meso-tetra(4-carboxyphenyl) porphine.

Tumor Tissue Culture Uptake Studies

P815 mouse leukemia tumor cells were grown for several days before use in RMPI 1640 tissue culture medium. Cell cultures were diluted 1:10 with fresh culture medium before uptake studies. Five ml portions of cell suspensions were dispensed into 60 mm x 15 mm tissue culture dishes. After 1 hr of incubation, 0.5 ml of ^{59}Fe-labelled hemin (specific activity 10 μCi/mg hemin; hemin concentration 0.5 mg/dish)

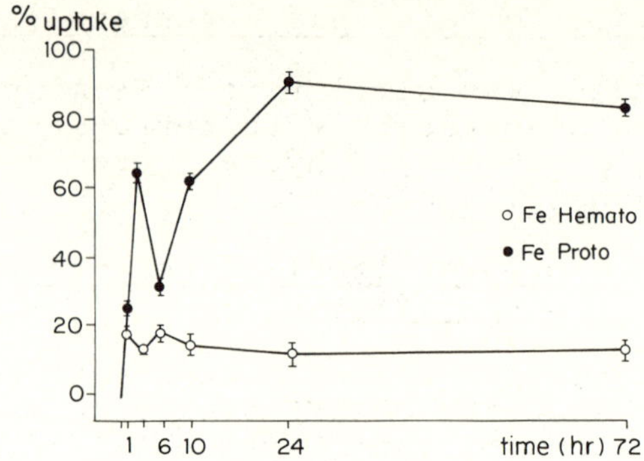

Figure 1. Tumor tissue culture uptake of previously labelled natural porphyrins.

or radionuclide was added to all but two dishes. A 0.5 ml sample of cell suspension at the start and end of the experiment was taken and radioactivity determined. At selected time intervals (0, 1, 3, 6, 12, 24, 48, 72 hr) a 0.5 ml sample was taken from each dish and filtered through a millipore filter. The cells were washed with 5 ml of PBS and counted for radioactivity. At time intervals noted above, 0.5 ml samples were also taken from the two non-radioactive tissue culture dishes and cell number determined with a Coulter Counter. All results were corrected for variations in level of radioactivity and cell number. Uptake at each time point was measured in duplicate, and the entire experiment was repeated five times.

Tumor Scanning

Tumor bearing rats with breast carcinoma on neck or side regions or with prostate carcinoma were injected with ^{52}Fe-chloride(III), ^{52}Fe-protohemin, ^{52}Fe-TMPI or ^{52}Fe-TCP (IV tail vein <5 mg porphyrin/animal, <0.5 ml). The animals were scanned 3 and 24 hr after injections.

RESULTS AND DISCUSSION

Tumor Tissue Culture Uptake Studies

The majority of hemins and some of the radionuclides showed an initial uptake followed by a decrease in activity. This was followed again by an uptake phase usually higher than the first.

Of the two radioactive labelled porphyrins used by other investigators, protohemin showed better uptake than hematohemin (Fig. 1). Hematohemin uptake was 18.3% of total

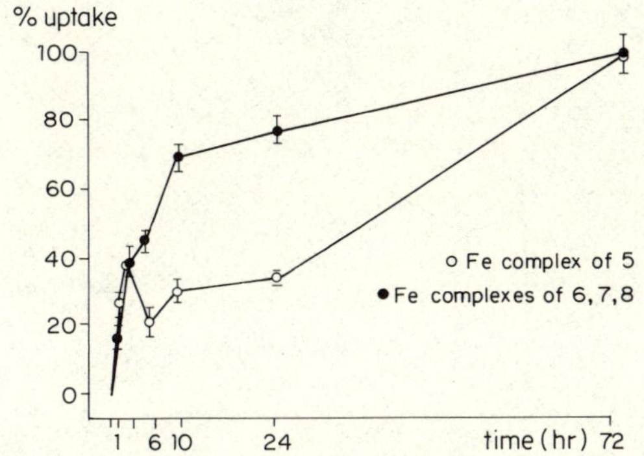

Figure 2. Tumor tissue culture uptake of natural hemins.

radioactivity at 1 hr, dropping to 10.6% at 24 hr and slowly increasing to 12.4% at 72 hr. Protohemin had a very high initial uptake (63.7%) at 3 hr followed by a decline to 30% at 6 hr and another rapid uptake phase to 87% at 24 hr.

Of the natural porphyrins used, deuterohemin derivatives showed the highest degree of early uptake (Fig. 2). The deuterohemin derivatives showed rapid uptake (69%) at 10 hr followed by a slow increase in uptake to 100% at 72 hr. The 2-formyl-4-vinyl, 2-vinyl-4-formyl, and the 2,4-diformyl deuterohemins had similar uptake values and are shown as one curve. Photoprotohemin showed a rapid increase to 37% at 3 hr; this was followed by a decline phase to 20% at 10 hr, and another increase to 100% at 72 hr.

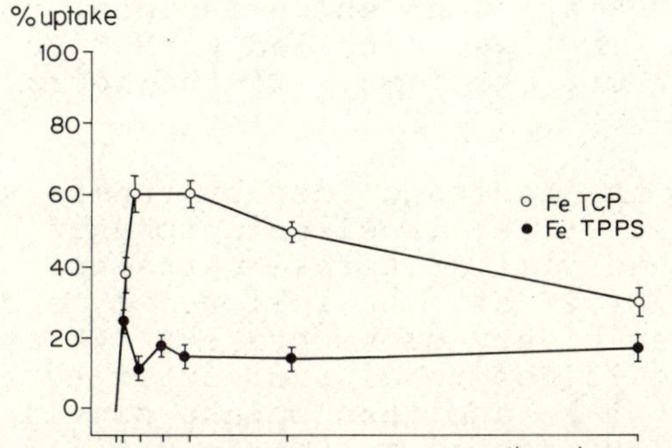

Figure 3. Tumor tissue culture uptake of artificial hemins.

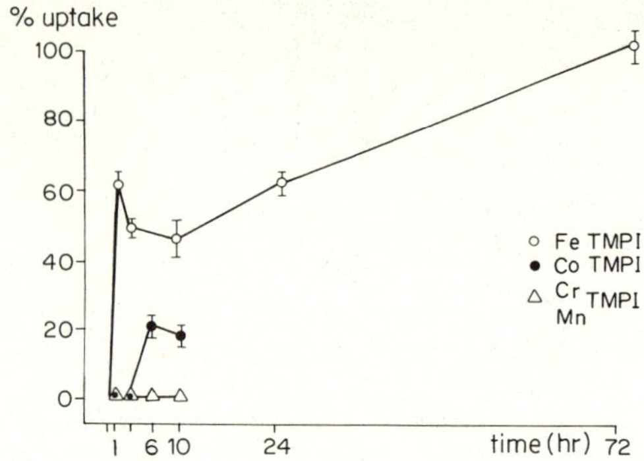

Figure 4. Tumor tissue culture uptake of TMPI labelled with different metals.

Of the artificial hemins tested, meso-tetra (4-carboxyphenyl) hemin (CP) showed very rapid uptake (60%) at 3 hr, and a loss (to 30%) at 72 hr (Fig. 3). Tetra-Na-meso-tetra (4 sulfonatophenyl) hemin (TPPS) had a low uptake curve as did hematoporphyrin. Meso-tetra (4-N-methylpyridyl) hemin tetraiodide (TMPI) also showed very rapid initial uptake at TPPS (60% at 1 hr); this decreased to 42% at 10 hr and increased to 100% at 72 hr (Fig. 4).

Because of the short half life of ^{52}Fe, only hemins with rapid initial uptake were further examined as potential scanning agents. These included protohemin, TCP, and TMPI. Other hemins that showed good uptake at 24 to 72 hr are not ideal due to the short half-lives of the radionuclides used for scanning.

If other radionuclides were to be used, TMPI would be the best choice since it shows very high initial uptake which is ideal for short-lived radionuclides; uptake continued up to 100% at 72 hr which is ideal for longer-lived radionuclides. TMPI was labelled with ^{59}Fe (as described above), ^{56}Co, ^{51}Cr, and ^{54}Mn. Of the four, Fe-TMPI was the best (Fig. 4).

Uptake of the radionuclides by themselves was very different from that of the labelled porphyrin (Fig. 5). Both ^{59}Fe and ^{67}Ga had similar uptake patterns. ^{67}Ga showed maximal uptake of 2% at 1 hr followed by a decline to 0.3% at 3 hr and then a very slow uptake (1.4% at 72 hr). ^{59}Fe was taken up to 13% of total radioactivity at 1 hr; this fell to 2.1% at 3 hr and then uptake slowly increased to 25% after 72 hr. ^{54}Mn showed no uptake until 3 hr; this rose to 32% at 6 hr. ^{56}Co, and ^{51}Cr showed no uptake at 10 hr.

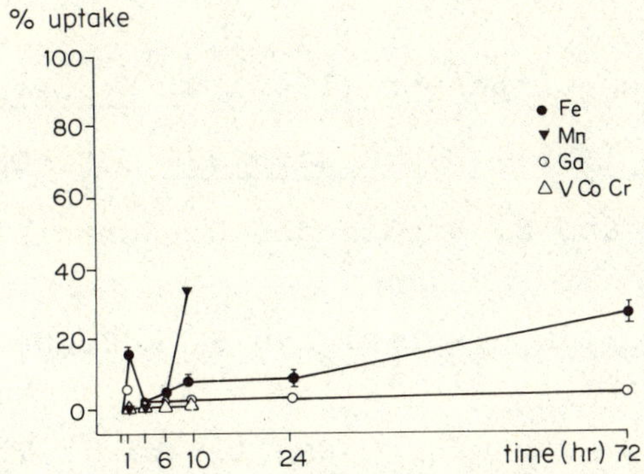

Figure 5. Tumor tissue culture uptake of various metals.

In Vivo Scanning

Studies using ^{52}Fe-TMPI (**10** as its iron complex), ^{52}Fe-TCP (**11** as its iron complex and ^{52}Fe-protohemin showed no tumor localization in rats at either 3 or 24 hr post injection. Most of the activity appeared in the liver which is in agreement with the report by Zanelli and Kaelin (1981). ^{52}Fe(III) chloride did accumulate in the tumor and RE system, but this may represent blood pool or localizing due to a transferrin receptor.

SUMMARY AND CONCLUSIONS

Fe-TMPI, Fe-TCP and protohemins showed high initial uptake using tissue culture techniques with mouse tumor cells (P815). However, the tumor uptake in the *in vivo* model was poor. Of the radionuclides investigated, iron labelled to TMPI demonstrated the best uptake by tumor *in vitro*. As has been previously demonstrated, the unnatural meso-arylporphyrin derivatives showed better uptake than did naturally occurring porphyrins and their derivatives.

ACKNOWLEDGMENTS

This work was supported by T.R.I.U.M.F. at the University of British Columbia, and the United States National Institutes of Health (GM 29198). We are grateful for the assistance of John Vincent and Hayes Dougan in the preparation of ^{52}Fe.

REFERENCES

1. Andwers, G.A., and Edwards, C.L. (1975) *J. Amer. Med.*

Assn 233, 1100.

2. Auler, H. and G. Banzer (1942) Z. Krebsforsch 53, 65.

3. Bases, R.E. (1954) BNL Report 326 Oct.-Dec., p. 45.

4. Bases, R.E. and S. Brodie and D. Rubenfeld (1967) Cancer 11,259.

5. Carrano, C.J., M. Tsutsui and S. McConnel (1977) Cancer Treatment Repts., 61, 1297.

6. DiNello, R.K. and C.K. Chang (1978) in The Porphyrins Ed. D. Dolphin, Academic Press, New York, Vol. I.

7. Falk, J.E. Porphyrins and Metalloporphyrins (1964), Elsevier, New York.

8. Figge, F. (1945) AAAS Research Conference on Cancer, Science Press, p. 117.

9. Figge, F.H.J., G.S. Weiland and L.O.J. Manganiello Proc. Soc. Exp. Biol. Med. 68, 640.

10. Kessel, D. (1981) Cancer Res. 41, 1318.

11. Manganiello, L.O.J. and F. Figge (1953) Bull. School Med. Univ. Maryland 38, 124.

12. Meyer-Betz, F. (1913) Deut. Arch. Skin Med. 112, 476.

13. New England Nuclear Gallium Citrate ^{67}Ga, Nuclear Department Reference, NEN, (1976) Mass.

14. Peck, G.C., H.P. Mack and F.H.J. Figge (1953) Bull. School Med. Univ. Maryland 38, 124.

15. Policard, A. (1924) Compt. Rend. Soc. Biol. 91, 1432.

16. Rassmussen-Taxdal, D.S., G.E. Ward and F.H.J. Figge (1955) Cancer Res., 8, 78.

17. Schwartz, S., K. Absolon and H. Vermund (1956) Univ. Minn. Med. Bull., 27, 7.

18. Winkelman, J. (1958) Cancer Res. 22, 589.

19. Winkelman, J. and J.E. Hayes (1963) Nature 30, 903.

20. Zanelli, G.D. and A.C. Kaelin (1981) Br. J. Rad. 54, 403.

EXCITED STATE INTERACTIONS OF PROTOPORPHYRIN IX AND RELATED PORPHYRINS WITH MOLECULAR OXYGEN IN SOLUTIONS AND IN ORGANIZED ASSEMBLIES

G. Sidney Cox and David G. Whitten

Department of Chemistry
University of North Carolina
Chapel Hill, North Carolina 27514

INTRODUCTION

We have studied the photooxidation of protoporphyrin IX and the related intermediates and products of its oxidation in homogeneous solution (mostly organic solvents) and in aqueous surfactant media, in particular, various micellar and vesicular media. Our contribution first discusses the reaction of protoporphyrin IX, the mono- and diformyl oxidation products, and the "green" hydroxyaldehydes formed from protoporphyrin IX. In solution, all of these compounds are good sensitizers for singlet oxygen; a small amount of the interaction of porphyrin excited states with molecular oxygen may result in electron transfer to give oxidized porphyrin (cation radicals) and superoxide anion. However, in most cases the singlet oxygen path is predominant. We have also measured the reactivity of the various porphyrin ground states with singlet oxygen. Our studies in organized assemblies show that reactivity may be quite different; studies still in progress are attempting to elucidate whether singlet oxygen remains the primary reactant in these processes. Preliminary results of quenching and sensitization studies in the organized media are discussed.

Porphyrins and metalloporphyrins have been demonstrated to be good sensitizers for photooxidations involving molecular oxygen (Cauzzo et al., 1977; Cannistraro et al., 1978). Protoporphyrin IX (**1**), the naturally occurring precursor to the hemes, has been found to sensitize its own

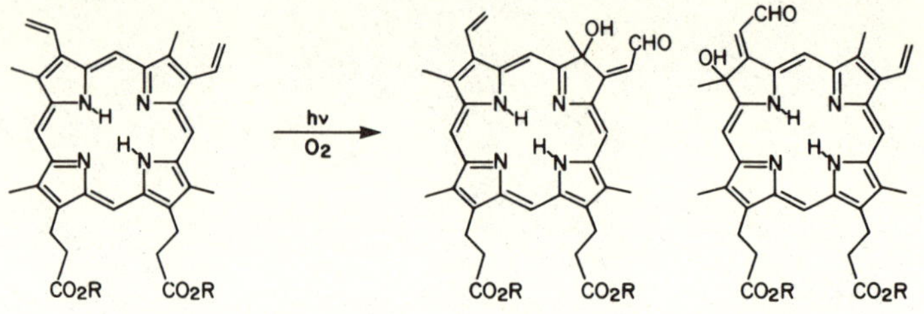

1 2 3 (1)

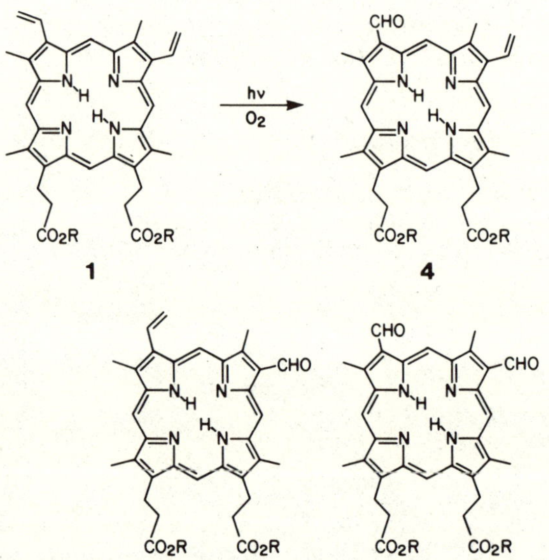

1 4 (2)

5 6

photooxidation (Fischer et al., 1938). While this reaction has been recognized for some time, the products have only been characterized in more recent studies. The major solution products, the green "photoprotoporphyrin IX", were shown to be isomeric hydroxyaldehydes (2) and (3) (eq. 1) (Inhoffen et al., 1969). Minor solution photooxidation products of 1 include the formyl derivatives **4**, **5** and **6** (eq. 2) (Horsey and Whitten, 1978; Cox and Whitten, 1981). We have studied the photoreactions of 1 and various ester derivatives in solution and in various organized media such as monolayer films, vesicles and micelles (Horsey and Whitten, 1978; Cox, 1981). These investigations have indicated that the product distribution is quite medium-dependent, with those products indicated in eq. 1 being predominant when the reaction is carried out in solution and those indicated in eq. 2 more prominent in the various organized assemblies. In this paper, we report results of some of our studies in this area which were undertaken to better elucidate mechanisms for photooxidation in various media, the specific sensitizing role of the porphyrins and the possible biological implications of the photooxidation of protoporphyrin IX and related compounds.

Most porphyrins and some metalloporphyrins are good sensitizers for the formation of singlet oxygen ($^1O_2$), the lowest excited state of molecular oxygen (eqs. 3,4) (Gollnick, 1968; Wayne, 1969; Wasserman and Murray, 1979). The reaction involves energy transfer from the excited triplet state of the porphyrin to ground state oxygen. This sequence of reactions is probably the most widely implicated of all photooxidation mechanisms. Following its generation singlet oxygen can attack a variety of substrates to generate oxygenated products. This mechanism can explain both sets of photoproducts obtained from protoporphyrin IX. Reaction of singlet oxygen with the diene units consisting of the vinyl group and one endocyclic double bond could lead to an endoperoxide (eq. 5) which could subsequently rearrange to give products **2** and **3**. Alternatively, 1,2 addition of singlet oxygen directly to the vinyl groups of **1** could lead to dioxetane which would be expected to cleave yielding a formyl product and formaldehyde (eq. 6) (Schaap and Zaklika, 1979).

$$P \xrightarrow{h\nu} P^{*1} \xrightarrow{ISC} P^{*3} \qquad (3)$$

$$P^{*3} + O_2 \longrightarrow P + {}^1O_2 \qquad (4)$$

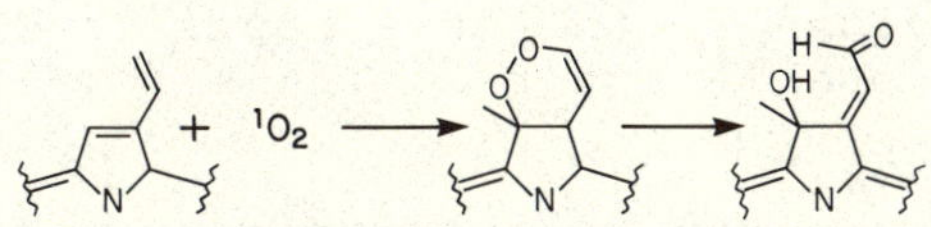

(5)

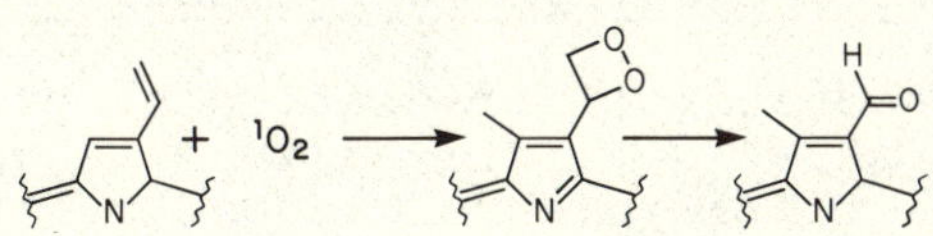

(6)

Recent investigations have suggested other possible mechanisms for photooxidations which might be expected to operate in the case of protoporphyrin IX (Spada and Foote, 1980; Smith, 1978; Kepka and Grossweiner, 1971). One possibility is that the excited porphyrin (either singlet or triplet might be operative in this case) could be quenched by one-electron transfer to give porphyrin cation-radical and superoxide anion (eq. 7) (Koizumi and Usui, 1972). These two radical ions could then recombine to yield products directly or undergo other reactions proceeding through various free radical intermediates. Although this type of reaction has not been considered an important pathway previously, recent investigations have indicated both that superoxide formation occurs on excitation of various dyes in the presence of oxygen and that photooxidations may often involve superoxide as an intermediate (Jahnke and Frenkel, 1978). In one case the product formed by the superoxide route was shown to be identical to that expected from singlet oxygen reaction (Eriksen et al., 1977).

$$P^{*} + O_2 \longrightarrow P^{+} + O_2^{\overline{\bullet}} \qquad (7)$$

Recently we examined the possibility that several porphyrins and their metal complexes could produce superoxide ion via eq. 7 by irradiating the various porphyrins in oxygenated solutions containing the spin trap 5,5'-dimethyl pyrroline-1-oxide (DMPO) (Harbour and Hair, 1978; Harbour and Bolton, 1977). We were able to detect superoxide formation from all of the porphyrins used in this study. While the efficiency of superoxide formation was found to be higher with several metalloporphyrins than for free base protoporphyrin IX, in all cases it was found that the formation of singlet oxygen was substantially more efficient than that for formation of superoxide (Cox et al., 1978). Subsequent investigations with several phthalocyanines and other species having lower triplet energies than **1** and the metalloporphyrins have demonstrated that for these pigments superoxide ion formation apparently does not occur even though singlet oxygen is efficiently produced (Maillard et al., 1980). Evidently in these cases reaction 7 is energetically unfavorable but reaction 4 remains exothermic.

MECHANISM OF PROTOPORPHYRIN IX PHOTOOXIDATION IN SOLUTION

As mentioned above, irradiation of protoporphyrin IX or several of its ester derivatives in organic solutions leads to the products shown in eqs. 1 and 2, with the hydroxyaldehydes **2** and **3** strongly predominating. Previous studies have demonstrated that protoporphyrin IX sensitized singlet oxygen and a quantum yield for its sensitization by **1** in carbon tetrachloride (ϕ=0.77) has been reported (Krasnovsky, 1979). We examined the photooxidation of the dimethyl ester of protoporphyrin IX (**1a**) and found that addition of typical singlet oxygen quenchers such as cholesterol and diazabicyclo(2.2.2)octane (DABCO) quenches its disappearance (Carlsson et al., 1972; 1974; Doleiden et al., 1974). DABCO, which is an excellent quencher of singlet oxygen but does not interact with superoxide in aprotic solvents or appear to affect the yield of superoxide produced in porphyrin sensitization of superoxide (Cox et al., 1978), sharply decreases the rate of formation of all of the photoproducts **2** - **6** from **1a** in methylene chloride (Cox and Whitten, 1981). Similar results were also observed in the more polar solvent N,N'-dimethylformamide with DABCO and in methylene chloride with cholesterol. The observation that the product ratios are not appreciably altered by addition of DABCO or cholesterol suggests that most, if not all, of the product formed in methylene chloride or DMF arises from a singlet oxygen mechanism.

$$P^{*} + MV^{2+} \longrightarrow P^{\dot{+}} + MV^{\dot{+}} \qquad (8)$$

$$MV^{\dot{+}} + O_2 \longrightarrow MV^{2+} + O_2^{-\cdot} \qquad (9)$$

$$P^{\dot{+}} + O_2^{-\cdot} \longrightarrow \text{Products} \qquad (10)$$

A slight increase in the ratio of formyl products to hydroxyaldehyde products at high (DABCO) suggests that a small amount of the former might be formed via an alternative pathway, perhaps involving superoxide. To test this possibility several experiments were carried out. Addition of di-t-butyl-4-methylphenol (DBMP) as a potential quencher resulted on only a very small decrease in the rate of photooxidation of **1a**. DBMP is an excellent radical scavenger which should efficiently quench superoxide (Jefford and Boschung, 1977); however the small effect observed does not provide any support of a superoxide mechanism in methylene chloride. A somewhat clearer indication of a possible role of superoxide in **1a** photooxidation was provided by studies using the electron acceptor N,N'-dimethyl-4,4'-bipyridine (methyl viologen, MV^{2+}) to direct the photochemical formation of porphyrin cation and superoxide anion. The viologen should quench excited states of **1a** by electron transfer (eq. 8) and subsequent reaction of the reduced viologen with molecular oxygen to generate superoxide (eq. 9) has already been demonstrated in other studies (Farrington et al., 1973; Harbour and Bolton, 1975). Subsequent reaction of the radical ions could thus lead to photooxidation products (eq. 10). Under conditions where eqs. 8 - 10 should operate (via quenching of the excited singlet state of **1a**) it was found that both the disappearance of **1a** and the hydroxyaldehyde product formation were strongly quenched. The formation of formyl products **4** - **6** was also quenched under these conditions, but not nearly so dramatically. Thus the results suggest that a superoxide path as outlined by eqs. 8 - 10 could result in selective formation of the formyl products **4** - **6**. The net result of the solution studies is, however, a clear indication that the major path for formation of all of the major products **2** - **6** is via singlet oxygen sensitization.

$$^1O_2^* \xrightarrow{k_d} O_2 \qquad (11)$$

$$^1O_2^* + P \xrightarrow{k_p} \alpha(2+3) + 1-\alpha(\text{Others}) \qquad (12)$$

$$^1O_2^* + Q \xrightarrow{k_q} Q + O_2 \text{ or } QO_2 \qquad (13)$$

The results of our study of the photooxidation of **1a** in solution suggest the mechanism given by eqs. 3, 4, 11-13. Here k_p is the total rate constant for quenching of singlet oxygen by **1a** and α is the fraction of the quenching events which leads to products **2** and **3**; k_q is the rate constant for radiationless decay of singlet oxygen. By monitoring the formation of products **2** and **3** which account for more than 90% of the products formed, it was possible to study the kinetics of the reaction and determine α and k_p. The value of $\alpha = 0.09$ indicates that over 90% of the interactions between singlet oxygen and **1a** result in only physical quenching. The total quenching constant for reaction 11 is 8.5×10^{-5} $M^{-1}sec^{-1}$; this value is moderately low compared to those obtained for quenching of singlet oxygen by several other porphyrins (Table 1) (Drasnovsky, 1979). Thus, even though the protoporphyrins possess potentially reactive vinyl groups, they are good sensitizers for singlet oxygen with relatively low probabilities of recapturing the singlet oxygen in solution, once sensitized.

We have also examined the photoreactivity of the products of **1a** photooxidation upon irradiation independently in aerated solutions. Table 2 indicates that all five of the products are good sensitizers of singlet oxygen with quantum yields ranging from 0.68 to 0.71. Each of the products are concurrently degraded on irradiation but with much lower efficiencies than protoporphyrin IX. Table 3 compares the reactivity of the different photoproducts with that of **1a**. Since the photodecompositions of products **2-5** can all be quenched by DABCO, their photodegradation must at least partially be due to singlet oxygen-induced photooxidations. The hydroxyaldehydes **2** and **3** are observed only to bleach upon irradiation in aerated solutions; no new products have been characterized or detected for irradiation of these photoproducts. In contrast, irradiation of **4** and **5** leads to new products having long-wavelength absorption characteristic of the hydroxyaldehydes **2** and **3**. Thus it appears that at least a major fraction of the reaction of

Table 1. Quenching Constants for Singlet Oxygen by Various Porphyrin Derivatives

Porphyrin	k_p ($M^{-1}s^{-1}$)
Chlorophyll a	7×10^8*
Tetraphenylbacteriochlorin	10^8*
Bacteriochlorophyll a	10^9*
Tetraphenylporphyrin	10^6*
Zinc Tetraphenylporphyrin	4×10^7*
Protoporphyrin IX DME	8×10^5**
2(4)-Formyl-4(2)-deuteroporphyrin	10^5***

*Data from Krasnovsky, A. A., Jr. (1979) Photoluminescence of Singlet Oxygen in Pigment Solutions, Photochem. Photobiol. 29, 29-36.
**Data from Cox, G. S. and Whitten, D. G. (1981) Mechanisms for the Photooxidation of Protoporphyrin IX in Solution, J. Am. Chem. Soc. in press.
***Data from Cox, G. S. (1981) unpublished results.

Table 2. Quantum Yields for Singlet Oxygen Production.*

Porphyrin		Quantum Yield
$\phi_{(1)}$	Protoporphyrin IX	0.77**
$\phi_{(2+3)}$	Photoprotoporphyrin IX	0.69
$\phi_{(4+5)}$	Monoformylmonovinyl-deuteroporphyrin IX	0.71
$\phi_{(6)}$	Diformyldeuteroporphyrin IX	0.68

*Based on relative rates for photooxidation of diphenylfuran in methylene chloride.
**Obtained from literature (Krasnovsky, 1979). Other values calculated relative to ϕ for 7.

these monoformyl porphyrins involves hydroxyaldehyde formation to give "mixed" oxidation products (eq. 14).

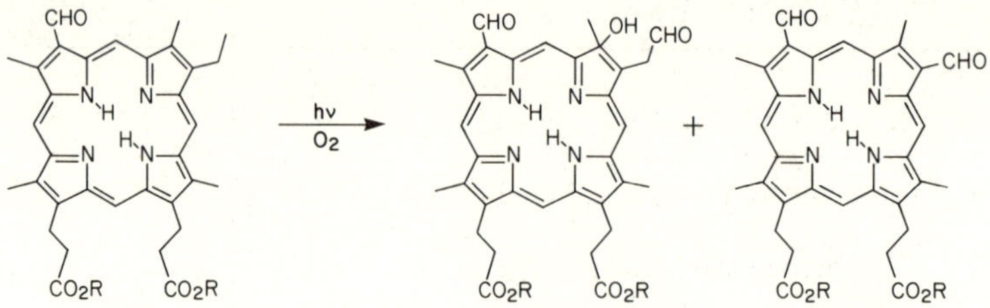

4 **7** **6** (14)

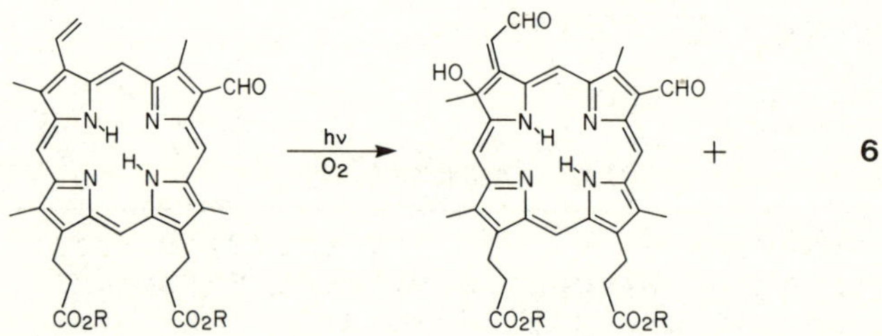

5 **8**

Product **6** is evidently also produced from irradiation of **4** and **5**, although in lesser amounts. We are currently investigating the parameters α and k_{ρ} for reaction of **4** and **5** with singlet oxygen; however the observation that the overall quantum yield for reaction of these monoformyl derivatives is only 12% that for **1a** suggests that these molecules are even poorer substrates for singlet oxygen than **1a**. This would be in line with the lower reactivity indicated for more oxidized porphyrins indicated in Table 1. The finding that all of the oxidation products of **1a** are more stable than **1a** but at the same time excellent sensitizers for singlet oxygen suggests that these products may play a major role in porphyria and other light-induced disorders related to porphyrin photooxidation.

PHOTOOXIDATION OF PROTOPORPHYRIN IX DERIVATIVES IN ORGANIZED MEDIA

As discussed earlier, preliminary studies on the irradiation of protoporphyrin IX derivatives in organized media such as monolayer films, supported multilayers and micelles resulted in pronounced differences in the product distribution of oxidation products **2** - **6** in these media compared to organic solutions. Since these media approximate to varying extents the lipophilic environments anticipated

Table 3. Relative Quantum Yields for Photodegradation of Protoporphyrin IX and Photoproducts (**2** - **6**).*

Porphyrin	$\frac{\phi \text{ Photoproduct}}{\phi \text{ Protoporphyrin IX}}$
Monoformylmonovinyldeuteroporphyrin IX (**4** + **5**)	0.12
Photoprotoporphyrin IX (**2** + **3**)	0.092
Diformyldeuteroporphyrin IX (**6**)	0.028

*8.0 x 10^{-4} M porphyrin in methylene chloride.

for protoporphyrin IX and many other relatively hydrophobic porphyrins in biological systems, it appears of particular interest to determine the rates and mechanisms for photoreactions of these porphyrins with molecular oxygen in such environments. Our work in this area is not yet complete; however, at the present time some data are available which underline some of the similarities and differences observed between solution and the various organized assemblies.

It is necessary to use protoporphyrin IX as well as various esters such as **1a**, **1b** and **1c** to obtain reasonable concentrations of porphyrins in the different media. As data in Table 4 indicate, rates of reaction, as measured by the disappearance of porphyrin for the different protoporphyrin IX derivatives, in fluid organized media are not dramatically different from that measured in methylene chloride. However, the rate of appearance of hydroxyaldehyde products **2** and **3** is much reduced. This is indicated in the product distributions listed in Table 5; thus in each of the media the yield of 1,4 addition products accounts for no more than 50% of the total observed product.

The reason for the change in product distribution on going from solution to organized media is at this time still unclear. From the solution studies it would appear possible that different oxygen intermediates could account at least in part for the difference in products. However, it does not seem too probable that such a change in medium should alter dramatically the mode of oxygen quenching of the excited porphyrin. Furthermore, preliminary studies indicate

Table 4. Relative Quantum Yield for **1** Photooxidation in Organized Media

Media	ø(1) Organized Media[1,2] / ø(1) Methylene Chloride	ø(2+3) Organized Media[2,3] / ø(2+3) Methylene Chloride
SDS[4]	0.76	0.27
Brij 35[5]	0.98	0.67
Vesicle[6]	1.39	0.83

[1]Relative quantum yields for disappearance of **1**.

[2]Ratios are calculated at identical concentration for

[3]Relative quantum yields for appearance of products **2** + **3**. organized media and methylene chloride solution; 1.4×10^{-5} M.

[4]Sodium dodecylsulfate; 1 porphyrin per 100 micelles.

[5]Polyoxyethylene (23) lauryl ether; 1 porphyrin per 110 micelles.

[6]L-α-Phosphatidyl choline dipalmitoyl;150 surfactant/ porphyrin.

that in monolayer assemblies, where the most drastic differences in product distribution compared to solution are observed, the photooxidation can be quenched by addition of cholesterol, a reagent expected to quench singlet oxygen but not superoxide. Studies are currently underway in the aqueous surfactant solutions using quenchers of both singlet oxygen and superoxide that would be expected to reside selectively in either the assembly, aqueous solution or at the solution-assembly interface. In very recent experiments we find that DABCO and azide ion both quench the photoreaction in neutral micelles, but not to a major extent. While a number of explanations could be advanced for this result, a reasonable possibility appears to be that only singlet oxygen escaping the assembly is quenched by these reagents. A plausible explanation for the change in product distribution is that singlet oxygen is involved in the major photoreactions occurring in all the different media but that the microenvironment influences considerable the relative occurrence of 1,4 vs 1,2 addition. The involvement of superoxide cannot be discounted, however, and given the

Table 5. Ratio of Photooxidation products in Organized Media

Media	1,4-Addition Products (2+3)	1,2-Addition Products (4,5,6)
Solution	9	1
Monolayer Films	1	2
Monolayer Assemblies	1	14
Micelles*	1	1
Vesicles**	1	1

*Sodium dodecyl sulfate and Brij 35.

**L-α-Phosphatidyl Choline Dipalmitoyl.

relatively polar environment provided by the hydrophobic-hydrophilic interface where the porphyrin generally should reside (Whitten et al., 1981), it is entirely reasonable that radical reactions emanating from species such as HOO˙ may be of much greater importance in these media.

The propensity of **1** and its initial photoproducts to form the relatively stable formyl products **4**, **5** and **6** in media resembling biological systems -especially phospholipid vesicles -points out that much of the sensitization of oxidation proceses may actually come form the formyl products such as **6** which are excellent sensitizers and extremely unreactive towards singlet oxygen.

ACKNOWLEDGEMENTS

We are grateful to the US Army Research Office (Contract DAAG29-80-K-0066) for support of this research. We also thank Ms. Christiane Bobillier for contributions in the studies of the photoproduct photooxidations.

REFERENCES

1. Cannistraro, S., A. V. DeVorst and G. Jori (1978) EPR Studies on Singlet Oxygen Production by Porphyrins, Photochem. Photobiol. 28, 257-259.

2. Carlsson, D. J., G. D. Mendenhall, T. Suprunchuk and D. M. Wiles (1972) Singlet Oxygen Quenching in the Liquid Phase by Metal (II) Chelates, J. Am. Chem. Soc. 94,8960-8962.

3. Carlsson, D.J., T. Suprunchuk, and D. M. Wiles (1974) The Quenching of Singlet Oxygen by Transition Metal Chelates, Can. J. Chem. 52, 3728-3737.

4. Cauzzo, G., G. Gennari, G. Jori and J. D. Spikes (1977) The Effect of Chemical Structure on the Photosensitizing Efficiencies of Porphyrins, Photochem. Photobiol. 25, 389-395.

5. Cox, G. S., D. G. Whitten and C. Giannotti (1979) Interaction of Porphyrin and Metalloporphyrin Excited States with Molecular Oxygen. Energy-Transfer versus Electron-Transfer Quenching Mechanisms in Photooxidations, Chem. Phys. Lett. 67, 511-515.

6. Cox, G. S. and D. G. Whitten (1981) Mechanisms for the Photooxidation of Protoporphyrin IX in Solution, J. Am. Chem. Soc. in press.

7. Cox, G. S. (1981) unpublished results.

8. Doleiden, F. H., S. R. Fahrenholtz, A. A. Lamola and A. M. Trozzolo (1974) Reactivity of Cholesterol and some Fatty Acids Towards Singlet Oxygen, Photochem. Photobiol. 20, 519-521.

9. Eriksen, J., C. S. Foote and T. L. Parker (1977) Photosensitized Oxygenation of Alkenes and Sulfides via Non-Singlet Oxygen Mechanism, J. Am. Chem. Soc. 99, 6455-6456.

10. Farrington, J. A., M. Ebert, E. J. Land and K. Fletcher (1973) Bipyridylium Quaternary Salts and Related Compounds. V. Pulse Radiolysis Studies of the Reaction of Paraquat Radical with Oxygen. Implications for the Mode of Action of Bipyridyl Herbicides, Biochim. Biophys. Acta 314,372-381.

11. Fischer, H. and K. Herrle (1938) Action of Light on Porphyrins. I. Transformation of Etioporphyrin I into Bilirubiniod Pigments, Hoppe-Seyler's Z. Physiol. Chem. 251, 85-96.

12. Gollnick, K. (1968) Type II Photooxygenation Reactions in Solutions, Adv. Photochem. 6, 1-122.

13. Harbour, J. R. and J. R. Bolton (1975) Superoxide Formation in Spinach Chloroplasts: ESR Detection by Spin

Trapping, Biochem. Biophys. Res. Commun. 64, 803-807.

14. Harbour, J. R. and M. L. Hair (1978) Detection of Superoxide Ions in Nonaqueous Media. Generation of Photolysis of Pigment Dispersions. J. Phys. Chem. 82, 1397-1399.

15. Horsey, B.E. and D. G. Whitten (1978) Environmental Effects on Photochemical Reactions: Contrasts in the Photooxidation Behavior of Protoporphyrin IX in Solution, Monolayer Films, Organized Monolayer Assemblies and Micelles, J. Am. Chem. Soc. 100, 1293-1295

16. Inholffen, H. H., H. Brockmann, Jr. and K. Bliesener (1969) Further Knowledge of Chlorophyll and Hemin, XXX. Photoprotoporphyrins and their Transformation in to Spirographis- and Isospirographis-porphyrin, Justus Liebigs Ann. Chem. 730, 173-185

17. Jahnke, L. S. and A. W. Frenkel (1978) Photooxidation of Epenephrine Sensitized by Methylene Blue--Evidence for the Involvement of Singlet Oxygen and of Superoxide, Photochem. Photobiol. 28,517-523.

18. Jefford, C. W. and A. F. Boschung (1977) The Reaction of Biadamantylidene with Singlet Oxygen in the Presence of Dyes, Helv. Chim. Acta 60, 2673-2685.

19. Kepka, A. G. and L. I. Grossweiner (1971) Photodynamic Oxidation of Iodide Ion and Aromatic Amino Acids by Eosin, Photochem. Photobiol. 14, 621-639.

20. Koszume, M. and Y. Usui (1972) Fundamental Aspects of the Oxidative and Reductive Photobleaching of Xanthene and Thiazine Dyes, Mol. Photochem. 4, 57-92.

21. Krasnovsky, A. A., Jr. (1979) Photoluminescence of Singlet Oxygen in Pigment Solutions, Photochem. Photobiol. 29, 29-36.

22. McCormick, J. P. and T. Thomason (1978) Near-UV Photooxidation of Tryptophan. Proof of Formation of Superoxide Ion. J. Am. Chem. Soc. 100, 312-313.

23. Maillard, P., P. Krausz, C. Giannotti and S. Gaspand (1980) Photoinduced Activation of Molecular Oxygen by Various Porphyrins, Bis-Porphyrins, Phthalocyanines, Pyridinoporphyrazins and their Metal Derivatives, J. Organomet. Chem. 197, 285-290.

24. Manring, L. E., J. Eriksen and C. S. Foote (1980) Electron-Transfer Photooxygenation. 4. Photooxygenation of trans-Stilbene Sensitized by Methylene Blue, J. Am. Chem. Soc. 102, 4275-4277.

25. Schaap, A. P. and K. A. Zaklika (1979) 1,2-Cycloaddition Reactions of Singlet Oxygen, In, "Singlet Oxygen" Wasserman, H. H. and R. W. Murray, Eds., First edition, Academic Press: New York.

26. Smith, G. J. (1978) Photo-oxidation of Tryptophan Sensitized by Methylene Blue J. Chem. Soc., Faraday Trans. 2, 1350-1354.

27. Spada, L. T. and C. S. Foote (1980) Electron-Transfer Photooxidation. 3. Detection of Radical-Ion Intermediates in the Cyanoaromatic Sensitized Photooxidation of trans- and cis-stilbene. J. Am. Chem. Soc. 102, 391-393.

28. Wasserman, H. H. and R. W. Murray (1979) "Singlet Oxygen" 1st ed.; Academic Press: New York.

29. Wayne R. P. (1969) Singlet Molecular Oxygen Adv. Photochem. 7,311-371.

30. Whitten, D. G., J. C. Russell, T. K. Forman, R.H. Schmehl, J. Bonilha, A. M. Braun and W. Sobol (1981) The Importance of Hydrophobic-Hydrophilic Factors in Reactions Occurring in Assemblies or at Interfaces: Studies Using Luminescent and Photochemically Reactive Probes, In, "Biometric Chemistry and Transition State Analogs: Approaches to Understanding Enzyme Catalysis," Green, B. S., Ed.; Elsevier Publishing co.: Amsterdam, in press.

INDEX